D1614362

TOMCAT ALLEY

NAVY

TOMCAT ALLEY

A Photographic Roll Call of the Grumman F-14 Tomcat

David F. Brown

Technical Editing by
Geoff LeBaron & Craig Kaston

Schiffer Military History
Atglen, PA

Acknowledgments

What kind of fool would want to produce a book featuring an image of every single Tomcat produced? This is exactly what I was thinking when I first decided to follow-up the best selling *Tomcats Forever*, co-authored with Robert F. Dorr nearly ten years ago. Many of my colleagues and fellow enthusiasts did not feel the concept was at all foolish. Instead, they, along with many past and present Tomcat pilots and maintainers, thought it would be great to possess a book containing an image of every Tomcat. A volume which would feature the various paint schemes, squadron markings, modifications and variations which have made the Tomcat what it is today, and will remain – the best fleet defender ever designed and manufactured.

I can honestly say work began on this project more than twenty years ago. I was attending an air show, at the insistence of my father. He wanted to show me a radical Navy fighter called the Tomcat. Since then I have attempted to gather as many images of and as much information about the Tomcat as humanly possible. My goal of acquiring an image of every F-14 would never be reached were it not for the assistance offered by my friends, colleagues, and fellow Tomcat enthusiasts around the world, many of whom spent hours searching their own Tomcat files for images and information. Therefore, I wish to extend a hearty thank you and well done to: Michael Anselmo, Dana Bell, Mike Boller, William F. Brabant, Sergio Bottaro, Bill Crimmins, Bill Curry, Robert F. Dorr, Lou Drendel, Robert Greby, Michael Grove, J.G. Handelman, Werner, Bob Henderson, Katie Hockenluber, E.S. Mule Holmberg, Alex Hrapunov, Phillip Huston, Marty Isham, Don Jay, Tom Kaminski, Craig Kaston, Mike Kopack Sr., G. Kromhout, Robert L. Lawson, Ray Leader, Geoff LeBaron, Don Linn, Don Logan, Pat Martin, Alfredo Maglione, Duncan McIntosh, Ron McNeil, Donald McGarry, Gerald McMasters, Joe Michaels/JEM Slides, Stephen Miller, Robert J. Mills, Al Mongeon, Rick Morgan, Dave Ostrowski, Kevin Patrick, Lionel N. Paul, Ron Picciani, Jeff Puzzullo, Yves Richard, Carl Richards, Brian C. Rogers, Angelo Romano, Barry E. Roop, Mick Roth, Bob Rys, Horace Sagnor Jr., John Sheets, John Shields, Douglas E. Slowiak, Keith Snyder, Don Spering, William D. Spidle, Bruce Stewart, Charles Stewart, Keith Svendsen, Charles E. Taylor, Regina and Robert Tourville, Bruce Trombecky, Jim Tunney, Scott Van Aken, Alan Vandam, Henk Van Der Lugt, Vance Vasquez, Peter Wilson, Jim Wooley, and Roberto Zambon. This project could not have been completed without your help.

In addition, I am grateful to those who assisted me in an official capacity: James Brooks, MSgt., USAF; Richard R. Burgess, LCdr., USN; Bobbie Carlton, CPO, USN; Daniel G. Carpenter, TSgt., USAF; Russell D. Egnor, USN Photo Division; ACE Ewers, PAO, NAS Oceana; J.C. Frier, Air Operations, NAS Oceana; Roy Grossnick and Todd Baker, Navy History Center; Jerry Leekey, Lt., USN; Chuck Mosely, Cdr., USN; David Rourk, PAO, USS Enterprise; James "Toadboy" Skarbek, Lt., USN; Troy Snead, PAO, NAS Oceana; Roger Seybel, Grumman History Center; and the U.S. Navy Safety Center.

Those of you who wish to remain anonymous, thanks, you know who you are. I apologize to anyone who I may have inadvertently overlooked. Finally, I wish to thank my sweetheart, my wife, for supporting me in this endeavor and for keeping the kids occupied, (sort of), while I worked on this project in my "tomb."

Dedication

Each of my previous aviation related pictorials have been dedicated to pilots, maintainers, or those who have made the ultimate sacrifice in defense of our freedom. Within the Tomcat community there are many individuals worthy of having this volume dedicated to them. Captain "Hank" Kleeman, who scored the Tomcat's first aerial victory, and Lieutenant Kara Hultgreen, the first female aviator assigned to an operational Tomcat squadron, quickly come to mind. Both made the ultimate sacrifice.

However, there are two individuals who really made this volume possible, my parents. Therefore, I am respectfully dedicating it to, Ruth and Francis "Bud" Brown. My mother ensured that I received a quality education and from early childhood my father nurtured my enthusiasm for military aviation.

Book Design by Robert Biondi.
Front dust jacket image courtesy of Rick Llinares and Chuck Lloyd,
Dash 2 Aviation Photography.

Library of Congress Catalog Number: 97-81278.

Printed in the United States of America.
ISBN: 0-7643-0477-1

We are interested in hearing from authors with book ideas on related topics.

Published by Schiffer Publishing Ltd.
4880 Lower Valley Road
Atglen, PA 19310
Phone: (610) 593-1777
FAX: (610) 593-2002
E-mail: Schifferbk@aol.com.
Please write for a free catalog.
This book may be purchased from the publisher.
Please include $3.95 postage.
Try your bookstore first.

Introduction

The text and more than seven hundred photos which follow will hopefully tell the Tomcat story bureau number by bureau number. I hope aviation enthusiasts, Tomcat aerophiles, maintainers and pilots past, present and future, find it interesting and rich in F-14 myth, legend and history. It is by no means intended as the definitive volume dealing with the subject. That goal will be left for another author and another time.

Nearly thirty years after its first flight, on 21 December 1970, the F-14 Tomcat continues to soldier on as the U.S. Navy's principal fleet defender. As of September 1997 there were thirteen deployable squadrons equipped with three variants of Grumman's Tomcat. In addition, the Tomcat equips one Fleet Replacement Squadron (FRS), a number of test and evaluation squadrons which continue to explore the type's full potential, and the U.S. Navy's Fighter Weapons School, better known as TOPGUN. A total of 712 Tomcats were manufactured between December 1970 and July 1992. Of this total, 79 were delivered to the Imperial Iranian Air Force prior to that country's Islamic revolution.

In order to examine the F-14 Tomcat even briefly, we must first look at a little known design of the 1950s, the F-6D-1 Missileer. As the name implies, the Missileer was designed to launch missiles – the Bendix XAA-M-10 air-to-air Eagle to be precise. The F-6D-1 showed promise in allaying U.S. Navy concern over the long-range cruise missile threat. Basically, the Navy was interested in acquiring aircraft possessing the ability to engage targets at a range beyond that of current air-to-air missiles.

The Missileer possessed this capability and could also guide several missiles to different targets simultaneously. Although it showed promise, the Missileer was proposed at a time when the Department of Defense was more interested in multi-mission aircraft. As a result the program was canceled before ever leaving the drawing board. However, development of the Eagle missile continued. The program was transferred to the Hughes Aircraft Company and would eventually evolve into the Phoenix missile system.

The Tactical Fighter Experimental (TFX) program resurrected the concept of downing your opponent before he could reach effective firing range. The TFX was a response to the Defense Department's mandate requiring weapons systems be both justifiable and cost effective. Victims of this logic were the XB-70 Valkyrie, Skybolt air launched ballistic missile and the Navy's Fleet Air Defense Fighter (FADF). Robert MacNamara, then Secretary of Defense, decreed the Navy's FADF would be merged with the USAF search for a new tactical fighter. This resulted in the F-111A/B program.

General Dynamics became the prime contractor for the F-111. Grumman was selected to build the navalized version of the F-111 due to its legendary experience with the design and construction of fighters for the Navy. The navalized version, known as the F-111B, flew from Grumman's Calverton, New York facility for the first time in May 1965. The type immediately encountered problems. The most serious concerned engine compressor stalls experienced at various points in the F-111Bs performance envelope. Additionally, the F-111B was determined to be too heavy for carrier operations. The death knell for the F-111B was sounded on 4 March 1968 at a Senate Armed Services Committee meeting on Capitol Hill. When queried by Senator John Stennis concerning the performance of the F-111B, Vice Admiral Thomas Connolly had this to say: "All the thrust in Christendom couldn't make a fighter out of this airplane." Afterward Congress simply stopped funding the navalized F-111. Having anticipated such an outcome, Grumman and the Navy were already well along with a replacement design.

By combining experience gained by Grumman during the design, construction and flight testing of its own swing-wing fighter, the XF-10F Jaguar, and certain aspects of the F-111B, such as the AWG-9 radar, Phoenix missile system, Pratt & Whitney turbofan engines, variable-geometry wing and two-man crew, Grumman eventually arrived at design number 303E, one of over 6,000 configurations investigated. Design 303E incorporated many of the features we have come to associate with the F-14 Tomcat. These include the AWG-9/Phoenix missile system, provision for Sparrow and Sidewinder missiles, Vulcan cannon, glove vanes, tandem seating under one canopy and variable geometry wing. The most notable difference between Design 303E and today's Tomcat was the option to produce the F-14 with twin vertical stabilizers as opposed to one. The Tomcat's swing wing was anchored by a unique wing box which combined Electron-Beam welding and high strength Titanium to produce an extremely strong wing box which saved 900 lbs compared to the bolt-up steel structure used in the F-111. Based on performance projections and the 303E design mock-up, the Navy awarded Grumman a contract on 4 February 1969, with the first flight scheduled to take place on or before 3 January 1971.

The XF10F-1 Jaguar was Grumman's first attempt at building a "swing-wing" fighter. The design and construction of this fighter were instrumental to Grumman receiving the contract to build the F-111B. (Grumman)

The first Tomcat, BuNo 157980 was delivered 25 October 1970, well ahead of schedule. Within two months it was ready for flight testing. The first test flight took place on 21 December 1970. It consisted of two slow speed circuits of the airfield. The second flight, on 30 December 1970 consisted of fuel dumping and gear retraction tests. At the controls was project test pilot William Miller. Chief test pilot Robert Smyth occupied the second seat. During the flight the Tomcat suffered a complete failure of both the primary and secondary hydraulic systems. As Miller attempted to land at Calverton, the last of the triple redundant hydraulic systems, the Combat Survival System, began to fail. Just prior to impact Miller and Smyth ejected. Neither suffered serious injuries. Although the Tomcat was totally destroyed, the crash did reveal the strength of the wing box. It was recovered from the crash site six feet under ground but very much intact. The cause of the crash was traced to vibration induced fatigue resulting in the total failure of the titanium hydraulic lines. The twelfth prototype, BuNo 157991 took the place of the number one prototype and flight tests commenced on 24 May 1971.

Flight testing proceeded with no major obstacles. Those minor problems which were encountered, rudder oscillation, spoiler buffet and intake buzz, were quickly corrected. Flight tests involved nineteen Tomcats flying from various locales, primarily Calverton, the Naval Missile Center (later Pacific Missile Test Center) at Point Mugu, and Naval Air Test Center at NAS Patuxent River. Carrier qualifications were conducted aboard the USS *Forrestal* (CV-59). By the summer of 1972, less than two years after its first flight, the Tomcat was deemed ready for its operational deployment.

The *Gunfighters* of VF-124, the west coast Fleet Replacement Squadron, received their first Tomcats in June 1972. The unit was tasked with training the first two operational F-14 squadrons, VF-1 and VF-2. The *Wolfpack* of VF-1 took delivery of their first Tomcat on 30 June 1973 at NAS Miramar. The *Wolfpack* and the *Bounty Hunters* of VF-2 took the type on its first combat cruise in January 1975, when they deployed aboard the USS *Enterprise* (CVN-65) to the Western Pacific. Although neither squadron fired their weapons in anger, they did provide valuable support for *Operation Frequent Wind*, the evacuation of Saigon, and received a Meritorious Unit Commendation. Meanwhile, a pair of Atlantic Fleet squadrons, the *Tophatters* of VF-14 and the *Swordsmen* of VF-32, were undergoing F-14 transition training with VF-124 at NAS Miramar. Old and new squadrons continued to transition to the Tomcat, reaching a peak of twenty-eight squadrons, including four reserve units, during the heyday of the Reagan/Bush administration. The current drawdown (read meltdown) has reduced the number of deployable squadrons to thirteen.

When the Shah of Iran grew concerned about overflights by Russian MiG-25s, Grumman lost little time promoting the Tomcat as a solution to this threat. It was first shown at the 1973 Paris Air Show, then again for the Shah's visit to the United States. The effort resulted in an initial order for 30 Tomcats signed in December 1973. More orders followed totaling 80 Tomcats. The first was delivered in January 1976. The last example, BuNo 160378, remained at Grumman for modifications to its aerial refueling system. Following the overthrow of the Shah, it was placed in storage at AMARC, Davis-Monthan AFB. It was later resurrected and assigned to the Naval Weapons Test Squadron at NAWS Point Mugu as an NF-14A. During the Iran-Iraq war, it was reported that at least ten Iraqi fighters were downed by Tomcats, some by the Phoenix missile. The Tomcat was also utilized as an early warning platform, using its powerful AGW-9 radar to detect low-flying Iraqi fighters. Due to the arms embargo and losses via attrition and combat, it is hard to speculate just how many Iranian Tomcats remain airworthy. The best estimates place the figure at between ten and twenty.

Although the Tomcat was offered to, and rejected by, the air arms of Spain, Australia, Canada, and Great Britain, possibly the largest customer lost was the USAF. Looking for a replacement for its aging fleet of F-106 Delta Darts, the USAF tested the F-14 during the mid-1970s and a full scale mockup was constructed, complete with conformal fuel pallets and ADC markings. Instead, the Air Force opted to purchase more F-15 Eagles. When comparing the air defense capabilities of both, the purchase of the F-15 seemed foolish. The Phoenix missile system alone was far more capable than the Eagle's AIM-7 Sparrow. Studies indicated

170 Tomcats could provide the same level of Defense as 290 F-15 Eagles. General Daniel 'Chappie' James, the former Commander of NORAD, personally evaluated the Tomcat. He recommended that it be purchased by the USAF. Today it is the F-15 Eagle which guards the sky over the continental United States; the F-14 falling victim to interservice rivalry and politics.

With the capacity to attack six different targets while tracking twenty-four, the Tomcat of the 1980s was the best weapons system in the air. What differentiated it from all others was its AWG-9 radar and its Phoenix missile system. The Tomcat is the only aircraft possessing the capability of launching AIM-7, AIM-9, AIM-54 and AIM-120 air-to-air missiles. The Phoenix can target a surface-skimming cruise missile or it can soar to 100,000 feet to down a MiG-25 *Foxbat.* The Phoenix's supersonic speed, 25g turn ability, and built-in ECM make it almost impossible to elude.

Designed to track and target fighter-sized aircraft at ranges in excess of 100 nautical miles, the AWG-9, a Pulse Doppler radar, can do so without the interference created by ground clutter. It can track its prey using one of six modes of operation; Pulse Doppler Search, Range While Search, Track While Scan, Pulse Doppler Single Target Track, Pulse Search, and Pulse Single Target Track. Any mode can be tailored to a particular set of circumstances. With its array of armament options and targeting modes, a young Navy Lieutenant (jg) could become an ace on his or her very first combat sortie.

For much of its service life, the Tomcat's operations have consisted mainly of routine deployments, with the exception of those to the Mediterranean Sea and Indian Ocean. Tomcats were on hand in April of 1980 to provide top cover during the failed attempt to rescue hostages held by Iran. The first real challenge to U.S. Navy Tomcats and its weapons system occurred during August 1981 in the Mediterranean Sea. On 19 August, a pair of VF-41 Tomcats, 'AJ/107', BuNo 160390 and 'AJ/102', BuNo 160403, were fired upon by Libyan Su-22 *Fitters.* The resulting engagement lasted less than one minute. The final tally was Tomcats 2 – Fitters 0.

During the remainder of the 1980s, the Tomcat maintained a constant vigilance in the middle east. Commencing in September 1982 and lasting through much of 1983, F-14s from squadrons like the *'Pukin' Dogs'* of VF-143 formed part of a multi-national Peace Keeping Force. Overflights of the hostile environs of Beirut and the Chouf Mountains of Lebanon were daily events. On more than one occasion Tomcats drew fire while gathering information on artillery positions and troop movements. Reconnaissance gathering Tomcats were equipped with the Tactical Air Reconnaissance Pod System, better known as the TARPS pod. Initially developed by the Naval Air Development Center for the A-7E Corsair II, the TARPS pod is carried on the Tomcat's number five Phoenix station. Planned as an interim replacement until the RF-18 joined the fleet, TARPS-modified Tomcats currently equip every operational F-14 squadron. Additionally, all F-14Ds are wired for TARPS. No longer a stopgap measure, the system is here to stay.

One of the most interesting uses of the Tomcat involved an air intercept over the Mediterranean Sea, in October 1985. Tomcats of VF-74, the *Bedevilers,* and VF-103, the *Sluggers*, were launched from the deck of the USS *Saratoga* (CV-60), to intercept an Egypt Air Boeing 737 which was known to be carrying four Arab terrorists. These men were responsible for killing an American citizen, Leon Klinghoffer, when they seized the Italian cruise ship *Achille Lauro.* The 737 was escorted to Sigonella, Sicily, where the terrorists were taken into custody.

One of the most controversial incidents involving aerial combat and the Tomcat took place on 4 January 1989. On this date, two VF-32 Tomcats, 'AC/207' BuNo 159610 and 'AC/202' BuNo 159437, were flying CAP for the USS *John F. Kennedy* (CV-67). While steaming away from Libyan waters, a pair of MiG-23 *Floggers* maneuvered into a collision course with the *Swordsmen* F-14s. With the F-14s almost within missile range of the MiG-23s the VF-32 flight leader initiated combat by launching a pair of AIM-7 Sparrows. Both missed their intended targets. A third AIM-7, launched by the wingman, destroyed a MiG-23. The flight leader

The F-111B was a direct result of the TFX program. It was designed to be the Navy's solution to defending the fleet against aerial threats including cruise missiles. It first flew on 18 May 1965, and with a staggering weight of 70,000 lb., it was soon obvious the design was ill-suited for carrier operations. (Grumman)

then fired a heat seeking Sidewinder which blew apart the second MiG-23. This entire aerial engagement lasted only eight minutes.

During *Operation Desert Storm*, it was the F-15 Eagle which received the bulk of the headlines for chalking up aerial victories. When this author interviewed F-15 drivers for *Birds of Prey*, many of them informed me of kills scored by F-15s after Iraqi aircraft had turned tail, away from the Tomcat's powerful AWG-9 radar. Only one aerial victory was claimed by a Tomcat crew during the Gulf War. This occurred on 6 February 1991, as Lt. Stuart Broce and Cdr. Ron McElraft were airborne in VF-1 Tomcat BuNo 162603. Tasked with providing CAP support for returning coalition strike aircraft, the flight was vectored to an aerial target by an Air Force E-3 Sentry. The crew observed a lone Iraqi Mi-8 *Hip* helicopter skimming across the desert. The crew launched a Sidewinder air-to-air missile which found and destroyed its target. One Tomcat, an F-14B, 161430, was downed by an Iraqi "Guideline" SAM on 21 January 1991. One crewmen was rescued, the other became a POW.

Since the end of the Gulf War, the Tomcat has played a key role in a number of peace keeping and contingency operations. These have included *Provide Comfort, Southern Watch, Provide Promise, Deny Flight*, and *Deliberate Force,* just to name a few. Missions consisted of armed reconnaissance, strike escort and BarCAP. When *Deny Flight* became *Deliberate Force,* the Tomcat demonstrated its ability to deliver precision guided munitions when they successfully bombed Bosnian-Serb targets.

Initially, the TF-30 engined F-14A was intended to be an interim model until a more reliable powerplant could be developed to match the aerodynamics of the highly advanced Tomcat design. In 1973, Grumman modified F-14A, BuNo 157986, with a pair of Pratt & Whitney F401-PW-400 engines. With the 30%+ increase in thrust, it was thought that this version, designated F-14B (not to be confused with the current F-14B), would become the definitive "Super Tomcat." However, the Tomcat program was now deeply overrun (partially due to inflation that would be the financial scourge of the 1970s), and Defense Secretary David Packard had cut the F401 engine (and other items such as the IRSTS) in a May 1971 Defense Systems Acquisition Review. The F-14B testbed flew a short test program and was mothballed for almost all of the next decade. The planned F-14C effort also died a premature death at this time. The F-14C would have had the F401 engines and a revised digital avionics suite giving the aircraft an all-weather air-to-ground attack and reconnaissance capability. The retention of the TF-30 would hobble the F-14 for years, but quite possibly helped to save the program from Congressional cancellation.

The next Tomcat variant appeared over a decade later when the Navy successfully proposed an F-14 that would include a set of General Electric F110-GE-400 engines as well as updated avionics and radar equipment. If a "Super Tomcat" does exist, then it is the F-14D. The latest and last production version of the F-14 Tomcat equipped with General Electric F110-GE-400 engines. These provide the F-14D with a thirty-two percent increase in thrust as well as carefree throttle handling, increased range and loiter time. The avionics suite was completely revised with digital avionics. These include the APG-71 Radar (reflecting fifteen years of advancement over the AWG-9), a new HUD and multi-function displays, Digital Stores Management Set, ASN-139 INS, ALR-67 RWR, belatedly the ALQ-165 ASPJ jammer, ASK-8 FEMS, and JTIDS data link; tied together by the Mil-Std-1553 Databus. The Databus connects the avionics in a sort of "Local Area Network" allowing standard and rapid data interface. Changing the external appearance of the aircraft were the Martin-Baker MB-14 Naval Aircrew Common Ejection Seats (NACES) and the return to the F-14 of an air-to-air Infrared Search and Track Set (IRSTS). The IRSTS is mounted in a dual chin pod with the Television Camera Set (TCS) carried on other Tomcats.

As initially planned, the F-14D was to have a production run of 127 and up to 400 of the earlier F-14As were to be converted. With the end of the Cold War these numbers were successively cut back. Faced with the competing financial demands of the Super Hornet (later the F/A-18E/F), the Naval Air tactical Fighter (NATF) and other competing funding demands, and despite vocal opposition by the Navy and some members of Congress, further F-14D production was canceled.

In the end, a total of 59 F-14D Super Tomcats were manufactured. This total consisted of 4 converted test aircraft, 37 new-built examples and 18 conversions from F-14As (12 by Grumman and 8 by NADEP Norfolk). The converted aircraft were designated F-14D(R) for remanufacture. The last F-14D and last new-built Tomcat, 164604, was accepted by the Navy on 10 July 1992. The F-14D currently equips three deployable squadrons, VF-2, VF-31, and VF-213. The Super Tomcat is also assigned to VF-101 and serves with several test and evaluation programs at Point Mugu and Patuxent River.

In the development chronology, the F-14D was followed by the F-14A+, (written officially "F-14A(PLUS)") even though it flew long before the F-14D. Since the development of the definitive F-14D avionics would take years, in order to save money the Navy had proposed to shut down the F-14 production line until the D was ready. Congress opted to keep the line open with a program to build F-14A+s with the older F-14A avionics and the F110 engine. Besides the improved powerplants, the airframes were equipped with limited avionics changes (inclusion of the ALR-67 RWR and ASK-8 FEMS), elimination of the glove vanes, and improved vents for the Vulcan cannon. A total of 70 aircraft (38 new and 32 converted F-14A) were procured, with another 15 F-14As being converted several years later as part of the Congressional compromise when the F-14D production effort was curtailed. On 1 May 1991, the Chief of Naval Operations ordered that all F-14A(PLUS) models be redesignated F-14B. This was done in part to eliminate a nonstandard aircraft designation and to avoid difficulties in the computerized logistics pipeline. In order to avoid further confusion, when I refer to the F-14B in this volume, I am making reference to the version delivered to the fleet, not to the engine testbed.

Today, while production has ended, Tomcat development has continued with several major efforts. Finally abandoning the "not a pound for air-to-ground" mentality, the Navy has decided to aggressively make use of the inherent capabilities of the aircraft, added software changes, "dumb" and Laser Guided Bombs to all aircraft and fitted the LANTIRN pod to selected aircraft. F-14 modification programs are extensive and are summarized as follows.

F-14B Upgrade

The "Upgrade" program updates the avionics in the F-14B aircraft with Mil-Std-1553 Databus, a hybrid Mission Computer with AYK-14 computer, Programmable Tactical Information Display

Grumman design, 303 was a direct outgrowth of the problems encountered with the F-111B program. When compared, the Grumman design was found to be superior to the F-111B. In May 1968 Congress put a stop to F-111B funding, effectively bringing the program to a halt. This mock-up is Grumman design 303-E. Of all the Grumman designs studied, it most closely resembles the Tomcat as we currently know it. (Grumman)

During the early 1970s, the USAF tested the F-14A to compare its suitability for the air defense role. This example was the Navy mockup modified for ADC by Grumman. Of interest are the conformal fuel tanks. The USAF selected the F-15 for the air defense mission and the project progressed no further. (Grumman)

(PTID), Programmable Multiple Display Indicator Group (PMDIG), and the AWG-15H stores management system which has been revised to support the new air-to-ground capability. The new digital architecture will allow future integration of the JSOW and JDAM stand-off air to ground weapons. Other modifications include wiring provisions to permit carriage of the ALQ-167 ECM pod and BOL Chaff Dispenser. F-14Bs are now being converted and emerge with a new designation painted above the Bureau Number: F-14B(Upgrade).

LANTIRN

Selected F-14A, F-14B and F-14D aircraft are being modified to carry the LANTIRN pod which was originally developed for the Air Force F-15E. A Navy/Contractor team performed a fast-paced integration program that saw less than eighteen months pass from inception to the first fielding of the system. LANTIRN-equipped Tomcats now have Infrared and Laser targeting capabilities that meet or exceed those of the F/A-18 Hornet. The system was formally unveiled on 14 June 1996, and VF-103 began a highly successful cruise with it aboard USS *Enterprise* (CVN-65), on 26 June 1996.

TARPS DI

New digital cameras are being studied to replace the film cameras in the once-interim Tactical Aircraft Reconnaissance Pod System. The new digital imaging system has the capability of transmitting images in 30 to 180 seconds at ranges up to 175 nm. TARPS DI is incorporated into the standard TARPS pod. The Pullnix digital camera is installed in the pod's first bay and replaces the wet-film KS-87 camera. Transmissions of digital images are sent via the Tomcat's UHF radio. TARPS DI was tested and put into service by the *Swordsmen* of VF-32. The squadron deployed with the TARPS DI, 25 November 1996 and the system is currently operational.

Global Positioning System

In accordance with DoD plans to have all military aircraft equipped with GPS by the year 2000, all Tomcats are being equipped as follows: F-14A and F-14B Upgrade aircraft are to be fitted with the EGI (Embedded GPS/INS) system which replaces the current mechanical gyro based Inertial Navigation System with a combined GPS receiver and Ring Laser Gyro INS. This will provide much greater precision and reliability. F-14D aircraft are being fitted with the MAGR (Miniaturized Airborne GPS Receiver).

In addition other efforts, such as night vision goggles and compatible cockpit lighting, and BOL chaff launchers are being fielded.

With these efforts, it can be seen that after the near disaster of program cancellation of the early 1990s, "The Cat is Back!" As this is written, F-14s are currently planned to fly until at least 2005. Since older F-14As have relatively little airframe life remaining the Navy is currently reducing their number and spreading the newer F-14B and F-14D models among the fleet squadrons. The current numbers are being reduced from 14 per squadron to 10. This reshuffle has freed enough of the newer Tomcats to permit VF-32 to re-equip with F-14Bs and VF-213 with F-14Ds. The remaining F-14A squadrons (VF-14, VF-41, VF-101, VF-154, VF-201, and VF-211) will continue to fly the type until successively replaced by the new F/A-18E/F. The new Hornet when delivered, will be in effect an entirely new airframe being only a passing resemblance to the F/A-18 that it will someday replace. Regardless of the fate of the existing Tomcats, it is safe to say that the venerable F-14 will continue to be the premier fleet defender and long range interdictor for the remainder of this decade and hopefully into the next century. **Tomcats forever!**

F-14 Roll Call

THE MISSION:

"TO INTERCEPT AND DESTROY ENEMY AIRCRAFT AND AIRBORNE MISSILES IN ALL WEATHER CONDITIONS IN ORDER TO ESTABLISH AND MAINTAIN AIR SUPERIORITY IN A DESIGNATED AREA. TO DELIVER AIR-TO-GROUND ORDNANCE ON TIME, IN ANY WEATHER CONDITIONS. AND, TO PROVIDE TACTICAL RECONNAISSANCE IMAGERY."

F-14A-01-GR, **157980**
Above: The first full scale development Tomcat, was photographed on 30 December 1970 while conducting fuel dumping and gear retraction tests. It crashed while on this flight due to a triple hydraulic failure. The crew consisting of William Miller and Robert Smythe ejected safely. (Grumman)

F-14A-05-GR, **157981**
Right: Delivered on 27 May 1971, the second FSD Tomcat was modified with retractable canards and a spin recovery chute. It was used to explore slow speed, high-lift, and stall/spin handling characteristics of the F-14. It suffered an in flight hydrazine APU fire on 13 May 1974. Pilot Chuck Sewell was able to land safely at Grumman's Calverton facility. Deemed beyond economical repair, the hulk was transported to NATF Lakehurst and it was stricken January 30 1976. (USN)

F-14A-10-GR, **157982**
Right: The third FSD Tomcat was delivered 4 January 1972 and was utilized in structural integrity tests. It is shown here exploring flight characteristics with an asymmetrical wing configuration. The Tomcat's wings could be swept a minimum of 20 degrees, to a maximum of 68 degrees. It was stricken 20 September 1994 and is currently on display at the Cradle of Aviation Museum at Mitchel Field, Long Island, NY. (Grumman)

F-14A-15-GR, **157983**
Left: Delivered on 26 October 1971 and transferred to Point Mugu, the fourth FSD Tomcat was used for testing of the AWG-9 fire control system. Following a vigorous test program, it was stored at AMARC commencing 27 October 1981. Later moved to NATTC Millington, TN, it was used to train firefighting crews. It was stricken on 20 December 1995. (U.S. Navy)

F-14A-20-GR, **157984**
Below: The fifth FSD Tomcat was delivered 3 December 1971 and transferred to Point Mugu on 12 December 1971. It conducted system installation test and compatibility work. The Secretary of the Navy, John H. Warner, arrived at NAS Miramar in this Tomcat on 14 October 1973, for formal ceremonies establishing the Navy's first two Tomcat squadrons, VF-1 and VF-2. It currently serves as a gate guard for NAS Pensacola (Bergagnini via Greby)

F-14A-25-GR, **157985**
Left: The sixth FSD Tomcat was delivered on 18 December 1971 and transferred to Point Mugu for missile separation and weapons tests. It shot itself down on 20 June 1972 when launching an AIM-7E-2, which pitched up rupturing and igniting several fuel cells. The crew ejected safely. The Tomcat impacted in the Pacific Ocean test range near Point Mugu. (Trombecky Collection)

F-14A-30-GR, **157986**
Right: Delivered on 19 September 1971, the seventh FSD Tomcat became the test bed for the Grumman F-14B. Not to be confused with the F-14Bs which are currently in the inventory, this version of the F-14B was equipped with Pratt & Whitney F401-PW-400 turbofan engines. The program was later canceled after only 33 flight hours. This airframe was later removed from storage and re-engined with GE F101 Derivative Fighter Engines(DFE) and later, the GE F110-400. This example is currently displayed aboard the USS Intrepid museum docked in New York City. (Grumman)

F-14A-35-GR, **157987**
Right: The eighth FSD Tomcat was delivered on 14 January 1972 and assigned to NATC Patuxent River for aerodynamic performance tests. It is viewed here on the NATC ramp loaded with Sidewinder and Phoenix missiles and external fuel tanks. It suffered an engine fire on the ground at NATC Patuxent River on 19 September 1974 and was stricken at NAS Norfolk on 27 April 1979. (Author's Collection)

F-14A-40-GR, **157988**
Below and opposite top: Following its delivery on 29 December 1971, this example was assigned to Point Mugu for evaluation of the AWG-9 radar and weapons system. Records indicate the ninth FSD Tomcat was stricken twice, once on 20 December 1985 and again on 16 September 1990. It currently guards the gate at NAS Oceana. (Hrapunov and Trombecky/Airframe Images)

F-14A-45-GR, **157989**
Left: Delivered on 29 February 1972, the tenth FSD Tomcat was assigned to NATC Patuxent River for initial carrier suitability tests. This Tomcat made the first carrier trap and launch aboard the USS Forrestal (CV-59) in June 1972. Test pilot Bob Miller was killed in this F-14 on 30 June 1972, when it crashed into the Chesapeake Bay while he was practicing for an air show. (Picciani Slides)

F-14A-50-GR, **157990**
Below: The eleventh FSD Tomcat was delivered 17 March 1972 and was initially assigned to Point Mugu for systems compatibility tests. It was photographed late in 1978 on the NATC ramp. This example was stored briefly at AMARC before being loaned to General Dynamics, Pamona, in 1985. It was stricken 31 December 1985 and is currently preserved at March AFB, CA. (Author's Collection)

F-14A-55-GR, **157991**
Right: Following the crash of 157980, this example was pulled off the line to assume the high speed test duties of the first FSD Tomcat. It displays the number one on its vertical stab but was officially designated 1X. It made the first Tomcat supersonic flight on 16 September 1971 and eventually reached a speed of Mach 2.41. Fitted with canards, it was assigned to NASA for spin and high-angle-of-attack tests. Upon its return to NATC, it was modified with external thrust diverters, leading the way to the X-31 enhanced maneuverability testbed fighter. This example was stricken on 16 September 1990. (Grumman)

F-14A-60-GR, **158612**
Above: The first production Tomcat was delivered on 12 May 1972. On 2 August 1972 it entered the anechoic chamber at Calverton for electromagnetic compatibility tests. When photographed in late 1978, it was assigned to NATC, Patuxent River. This Tomcat was rebuilt to block 135 standards, and assigned to VF-201 As of February 1998 it was the oldest active F-14A in the inventory. (Author's Collection)

F-14A-60-GR, **158613**
Right: Delivered on 9 June 1972, this Tomcat was initially used for maintenance and reliability tests in addition to carrier suitability tests following the loss of number 10. It was later loaned to NASA, and was utilized for laminar air flow studies. This example was stricken on 9 April 1992 and was transferred to the Naval Surface Weapons Center, Dahlgren, VA, to be used for target practice and ordnance testing. (Greby)

F-14A-60-GR, **158614** (TARPS)
Above: Delivered on 31 August 1971, this was the first F-14 to be assigned to pilot training. It was also the first Tomcat inducted into AMARC on 16 April 1980 receiving code 1K001. In 1985 it was ferried to Grumman to receive a block 135 upgrade and later became the earliest Tomcat to be modified for TARPS. Delivered to VF-202, it is pictured in that squadron's markings in May 1992. Following the disestablishment of VF-202, it was transferred to VF-201 in 1994. As of February 1998 it was still assigned to the *Hunters* of VF-201. (Grove)

F-14A-60-GR, **158615**
Left: Originally constructed as a Block 60, and delivered on 27 September 1972, this Tomcat served with VX-4 and was the first to wear new Naval Missile Center markings. Following rebuild to Block 135 standards, it served with two Naval Reserve squadrons prior to joining VF-14 in 1995. As of February 1998, it was one of the oldest Tomcats in the inventory. (U.S. Navy)

F-14A-60-GR, **158616**
Left: Delivered on 14 November 1972, the 17th Tomcat was used to replace number ten following its crash in June 1972. It took up 157989's regimen of carrier suitability tests. It was photographed 13 May 1973 at NATF Lakehurst. It was delivered to AMARC on 16 May 1980 and received storage code 1K002. Following rebuild to Block 135 standards, it was assigned to VF-201 of the Naval Reserve in 1989. As of February 1998 it was still serving with VF-201. (Spering)

F-14A-60-GR, **158617**
Right: Delivered on 6 October 1972, this Tomcat was photographed in 1977 while assigned to the Pacific Missile Test Center. This example was the first Tomcat to be flown by VF-124 instructors when it was temporarily loaned to the *Gunfighters* during October 1972. As of February 1998 it was assigned to VF-14 at NAS Oceana. (Bruce Trombecky/ Airframe Images)

F-14A-60-GR, **158618**
Below: The *Evaluators* of VX-4 were heavily involved in testing the Tomcat. This example, delivered on 31 October 1972, was photographed at Point Mugu in November 1973 while undergoing Operational Evaluation (OpEval). This F-14A arrived at AMARC on 2 June 1982 and was briefly stored there until being rebuilt as a Block 135. As of February 1998 it was assigned to VF-201. (Logan)

F-14A-60-GR, **158619**
Right: Obviously an air show shot, taken at NAS Patuxent River 15 September 1974. This example was utilized at NATC to test the prototype of the Pratt & Whitney TF30-P-414. A single example was installed along side a standard TF30. It entered a flat spin during testing and crashed near Patuxent River, on 22 February 1977. Despite this accident, the -414 was cleared for use by the Tomcat fleet. (Spering)

F-14A-65-GR, **158620** (TARPS)
Left: Following delivery on 31 December 1972, this Tomcat served with VF-124, the west coast Tomcat training squadron. It was transferred to VF-101, the Navy's east coast Tomcat training squadron, on 25 August 1976. One of twenty-six F-14As rebuilt to Block 135 standards, it was further modified to carry TARPS and assigned to VF-201. In September 1989 LCdr. Randy Adrian and LCdr Blake Estes escaped injury when the starboard LOX access door came loose and struck the canopy shattering 80% of the Plexiglas. Due to superior airmanship, this Tomcat made a safe landing at NAS Dallas. As of February 1998 this Tomcat was still assigned to VF-201. (Tunney)

F-14A-65-GR, **158621**
Left center: On 14 May 1973, Cdr James Taylor and Lt. Kurt Strauss of VF-124 flew from NAS Miramar to Patuxent River and took receipt of this Tomcat which at the time was assigned to NATC. Following a 21 May 1973 trans-Atlantic flight, they flew eleven shows of six to eight minutes duration each at the Paris Air Show from 25 May to 4 June. The show consisted of the following maneuvers: a half Cuban Eight at takeoff, going through the top at 2500 to 3000 feet with 200 to 250 knots airspeed; a slow roll, sweeping the wings fore and aft; knife edge pass at 350 to 400 knots; six-G, 360 degrees steep turn of 2000 foot radius at 500 feet; a tuck-under break to slow to landing speed, lowering the gear and extending the flaps; a wing walking maneuver at 105 knots in landing configuration down air show center; and finally, a near vertical climb, still with wings, flaps and gear down, on afterburner to landing pattern altitude, from which approach and landing was made. This example was later modified as an NF-14A, and was stricken 7 March 1995. (Spering)

F-14A-65-GR, **158622**
Below: This example was photographed shortly after its delivery on 24 March 1973. During December 1973 several USMC officers were in the process of being trained as F-14 instructors. The first men destined for USMC Tomcat squadrons arrived in June 1976. The training ended when the Marine Corps opted to remanufacture their F-4Bs to F-4Ns and their F-4Js to F-4S configuration. This example later served with PMTC and arrived at AMARC on 18 December 1991. It was stricken 17 October 1994. (Author's Collection)

F-14A-65-GR, **158623**

Right: A number of Tomcats were remanufactured, sometimes more than once. This example was rebuilt to Block 135 standards and was reassigned to VF-124 and then the PMTC. It was utilized for various air-to-air missile tests including the AIM-120 AMRAAM. It was photographed in flight with AIM-9 Sidewinders and AIM-54 Phoenix. Stricken on 23 September 1993, this example currently guards the gate at NAWS Point Mugu. (Vasquez)

F-14A-65-GR, **158624**

Above: Following service with VF-124 and VF-101, this F-14A was assigned test duties at NATC and PMTC. It later went to NADEP for upgrade to Block 135 standards. As of February 1998, it was assigned to VF-201. (Taylor)

F-14A-65-GR, **158625**

Right: This NF-14A was assigned to PMTC when it was photographed there on 4 December 1995. Of interest are the new Weapons Test Squadron tail markings. This Tomcat was delivered to VF-124 on 26 April 1974. It served with VF-101 prior to its transfer to PMTC in 1980. As of March 1997 it was assigned to the Naval Weapons Test Squadron at Point Mugu. (Vasquez)

F-14A-65-GR, **158626**
Above: Delivered on 2 May 1973 this NATC F-14A was photographed in October 1984. Modified to block 135 standards, it was assigned to VF-201. It was stricken 7 September 1990. (Author)

F-14A-65-GR, **158627**
Left: This Tomcat became the first F-14A delivered to VF-1 when Lt. C. D. Pentecost and LCdr. W. R. Mullins flew it from Calverton to NAS Miramar on 30 June 1973. It also served with VF-124 and VX-4. In 1983 it was delivered to NADEP Norfolk for rework to Block 135 standards. As of February 1998, it was assigned to the *Hunters* of VF-201. (Grove)

F-14A-65-GR, **158628**
Left and next page top: Photographs like these should send shivers down the spines of tailspotters everywhere. The starboard side of this VF-124 Tomcat displays BuNo. 158628. The portside, indicates this Tomcat is BuNo. 158627. Which is correct? Take your pick! As of September 1997, 158628 was assigned to VF-201. (LeBaron)

See previous entry.

F-14A-65-GR, **158629**
Right: This example was photographed in August 1987 departing Grumman's Bethpage, New York facility following a Block 135 upgrade. It was initially delivered on 27 June 1973 and served with VF-124 before its rebuild and delivery to VF-202. It was transferred to the only remaining Naval Reserve Tomcat squadron, VF-201 in 1994. It was still assigned there as of February 1998. (Kaminski)

F-14A-65-GR, **158630**
Below: The *Flying Checkmates* of VF-211 stood up as an F-14A squadron on 1 December 1975. This example, with its colorful markings, was photographed 17 November 1979 at NAS Miramar. During the mid-1980s it was modified to Block 135 standards and assigned to Naval Reserve Squadron VF-201. As of February 1998 it was still assigned to the *Hunters* of VF-201, based at NAS Fort Worth JRB. (Huston)

F-14A-65-GR, **158631**
Above: This beautifully posed Strike Test Directorate F-14A was captured on film at Patuxent River. This example was also rebuilt as a Block 135 and reassigned to VF-201. As of February 1998 this Tomcat was at NAS Jacksonville awaiting modifications, or disposition. (Paul)

F-14A-65-GR, **158632**
Left: The Naval Reserve goes to "the boat" on a regular basis for CarQuals. This F-14A of VF-201 was assigned to the USS Dwight D. Eisenhower (CVN-69), when photographed landing at NAS Dallas during July 1989. As of February 1998 this Tomcat was being overhauled at Grumman's St. Augustine facility. (Snyder)

F-14A-65-GR, **158633**
Below: Initially delivered to VF-124, this example was reassigned to VX-4 prior to its overhaul. Following Block 135 upgrades, this Tomcat joined VF-201 where it was assigned as of February 1998. Of interest is the chin mounted Television Camera Set which permits automatic and manual target acquisition. (Grove)

F-14A-65-GR, **158634**
Above: Delivered to VF-124 on 14 September 1973, this Tomcat was photographed 5 May 1974, a month after Capt. Lamoreaux gave General Daniel "Chappie" James an orientation flight in an F-14A. Although the former Commander of NORAD was duly impressed, interservice rivalries and politics prevented the acquisition of the Tomcat by the USAF. On 30 October 1986 this Tomcat was the first example delivered to VF-201. It has remained with the *Hunters* through February 1998. (Curry)

F-14A-65-GR, **158635**
Right: The contrast between the paint scheme on this VF-201 Tomcat and 158634 is quite striking. This scheme, known as the Tactical Paint Scheme, but often referred to as "low-vis", consisted of Medium Gray FS 35237 upper surfaces and Ghost Gray, FS 36320 on the lower surfaces. This author has seldom seen two low-vis Tomcats painted exactly the same. This example initially served with VF-124 and NATC prior to rebuild to Block 135. It was transferred to the Naval Reserve. As of February 1998 it was assigned to VF-14. (Snyder)

F-14A-65-GR, **158636**
Right: Delivered on 13 October 1973, this former VF-124 Tomcat was cocooned at NARF Norfolk awaiting a block 135 upgrade when it was photographed during June 1977. It was reassigned to VF-202, a Navy Reserve Squadron during June 1987. Following the disestablishment of VF-202 on 1 October 1994, this Tomcat was transferred to VF-201, another reserve unit. It was later assigned to VF-101. As of February 1998 it was on strength with VF-41. (Author's Collection)

F-14A-65-GR, **158637**(TARPS)
Left: When it arrived at NAS Miramar on 13 October 1973, VF-124 had just completed its first year of Tomcat operations. This F-14A was subsequently transferred to VF-101 on 5 November 1975. It was being utilized in the test role when photographed at NATC in November 1978. It subsequently was transferred to PMTC and later rebuilt to block 135 standards with TARPS capability. As of February 1998 it was assigned to VF-201. (Author's Collection)

F-14A-70-GR, **158978**(TARPS)
Below: Barry E. Roop, one of the best ground-to-air photographers in the business, caught this VX-4 F-14A, blasting off in zone 5 afterburner from McGuire AFB on 5 September 1988. This Tomcat introduced the standard wing glove fairing with shorter outboard wing fences. It was transferred to VF-1 in January 1992, and was stricken 18 February, 1993. It was briefly displayed at the NAS Miramar gate. (Roop)

F-14A-70-GR, **158979**
Left: This VF-1 Tomcat was painted in an experimental splinter scheme when it was photographed on 5 June 1976. The 100 modex signifies this *Wolfpack* F-14A is a "CAG Bird." Delivered in October 1973, this example served with VF-1 and then VF-301. It flew with the *Devil's Disciples* of VF-301 from 1984 until it was placed in storage at AMARC on 30 August 1990. It was stricken on 17 October 1994. (Logan)

F-14A-70-GR, **158980**
Above: Yet another attractive Tomcat paint scheme involved an overall gloss gull gray scheme as applied to this VF-302 F-14A. From its delivery date in 1973 until retirement at AMARC, 13 September 1990, this aircraft was assigned to only three NAS Miramar squadrons, VF-2, VF-124, and VF-302. It was stricken 17 October 1994. (Anselmo)

F-14A-70-GR, **158981**
Right: Once the fog burned off, there were a number of excellent locations at NAS Miramar to photograph Tomcats. This example, assigned to the *Stallions* of VF-302, a Naval Reserve Squadron, was photographed at Miramar in October 1991. Compare this low-vis tactical paint scheme with that of 158980 shown above. This Tomcat was stricken 4 June 1993 when it departed controlled flight during a 1 v 1 ACM sortie near San Clemente Island. The crew was rescued by helicopter. (Author)

PHOTO NOT AVAILABLE

F-14A-70-GR, **158982**
Right: This Tomcat was delivered on 11 December 1973. On 2 January, 1975 it was assigned to VF-1 and flying a local VFR Air Intercept Control training mission out of NAS Cubi Point, Republic of the Philippines. The crew, pilot LCdr. Grover Giles and NFO, LCdr. Roger McFillen, experienced a loud thump followed by loss of control and fire. Both crewmen ejected successfully.

F-14A-70-GR, **158983**
Left: Delivered on 23 November 1973, this F-14A was photographed on 22 May 1977 in the markings of the *Bounty Hunters* of VF-2. This Tomcat later served with VF-51 and VF-301. It was stricken on 20 June 1986, while assigned to VF-302 and conducting ACM training at NAS Fallon. The aircrew ejected successfully. (Leader)

F-14A-70-GR, **158984**
Left: Delivered 21 December 1973 this F-14A has flown with VF-1, VF-211, VF-111, VF-301, VF-302, VF-201, and VF-213. As of September 1997, it was the only remaining F-14A still operational from the 29 Block 70 Tomcats constructed. (Romano)

F-14A-70-GR, **158985**
Below: During the late 1970s, a handful of squadrons experimented with various camouflage schemes. This example, known as the Ferris or Splinter Scheme, was applied to VF-2's CAG bird. This Tomcat was stricken on 30 March 1992 following service with VF-2, VF-124, VF-1, and VF-191. (Logan)

F-14A-70-GR, **158986**
Above: When photographed in April 1996, this F-14A, formerly of VF-301, had already been in storage since 19 July 1991. Its history included service with VF-2 and VF-124. It was flown to AMARC in July 1991, and even though officially stricken on 21 February 1995 it was still in residence there in 1997. (Author)

F-14A-70-GR, **158987**
Right: This pre-delivery photo was taken December 1973 at the Grumman facility. The number 48 indicates this is the 48th Tomcat constructed. Assigned to VF-2, this F-14A spent its entire service life at NAS Miramar with VF-2, VF-124, and VF-301. It arrived at AMARC 25 October 1990. It was stricken 17 October 1994. (Paul)

F-14A-70-GR, **158988**
Rght: A 1944 *Life* magazine article described VF-2 as "the hottest Fighter Squadron in the Pacific." More than thirty years later VF-2 displayed some of the hottest markings ever applied to a Tomcat. This F-14A wears an overall gloss gull gray scheme, Langley Stripes and the *Bounty Hunters* insignia. Before reaching AMARC, this Tomcat would also serve with VF-124 and VF-301. Placed in storage on 18 September 1990 it was stricken on 17 October 1994. (Snyder)

F-14A-70-GR, **158989**
Left: This image is a classic view of the high-vis paint scheme. Delivered 21 December 1973 and following service with VF-1 and VF-124, this Tomcat spent the remainder of its career assigned to Naval Reserve Squadron, VF-302. It entered storage at AMARC, 29 August 1991 and was stricken 17 October 1994. (Van Geffen via Greby)

F-14A-70-GR, **158990**
Above: In early 1974 this F-14A was delivered to VF-1 and assigned code NK/103. It spent its entire operational service at NAS Miramar retiring from VF-301 in 1992. It was delivered to AMARC on 26 August 1991 and stricken on 17 October 1994. (Trombecky/Airframe Images)

F-14A-70-GR, **158991**
Left: The 52nd Tomcat goes through pre-delivery tests at Calverton. This example spent its entire operational career at NAS Miramar assigned to VF-1, VF-124, and VF-194. It was delivered to AMARC and assigned storage code 1K0011, on 22 August 1990. It was stricken 17 October 1994. (Paul)

F-14A-70-GR, **158992**
Right: The majority of the twenty-nine block 70 Tomcats were at one time or another assigned to VF-1 and/or VF-2. This VF-2 F-14A was photographed at NAS Miramar in 1974. The *Bounty Hunters* became the first Tomcat squadron to be awarded the CNO Naval Operations Safety Award. This example later served with VF-24, VF-51, and VF-301 before retiring to AMARC on 28 February 1991. Although stricken 17 October 1994, this Tomcat was still resting at that venue in August 1997. (Via Jay)

F-14A-70-GR, **158993**
Below: The *Wolfpack* of VF-1 was the first operational Tomcat squadron. It was also the first squadron, along with VF-2, to take the Tomcat on a combat cruise. VF-1 and VF-2 flew numerous CAP missions during *Operation Frequent Wind*, the April 1975 evacuation of Saigon. This Tomcat was accepted on 20 December 1973. It served with VF-1, VF-124, VF-51, and VF-302. It was flown to AMARC on 25 October 1990 and stricken on 17 October 1994. (Lawson)

F-14A-70-GR, **158994**
Right: Reserve Tomcat squadron VF-301, known as the *Devil's Disciples* are gone – disestablished 31 December 1994. The squadron will long be known for their safety record which covers more than twenty-four years and 71,322 flight hours without a class "A" mishap. This example was delivered to VF-2 on 17 May 1974. It served with VF-1, VF-124, and VF-301 before arriving at AMARC 27 August 1990. It was stricken 17 October 1994. (Slowiak)

F-14A-70-GR, **158995**
Left: This example was delivered to VF-2, early in 1974. It was reassigned to VF-124 and then VF-1. On 27 March 1978, it crashed onto a freeway while on final approach to NAS Miramar. The NFO, Ltjg. W.T. Laskowski was killed. The pilot survived but was injured. (Rys via LeBaron)

F-14A-70-GR, **158996**
Left: VF-1 has displayed some of the most eye-catching markings of any fighter squadron assigned to either coast. This example was photographed on 4 March 1974, just weeks after its delivery. While assigned to VF-2 it departed controlled flight on the downwind leg to NAS Miramar and crashed on 28 June 1977. (Author's Collection)

F-14A-70-GR, **158997**
Below: This VF-2 Tomcat prepares to launch from the USS Enterprise (CVN-65) during its first cruise in 1974-75. Of interest are the live AIM-7 and AIM-54 air-to-air missiles. This F-14A later served with the PMTC and completed its career with VF-301. It was stricken 4 February 1992. (USN)

F-14A-70-GR, **158998**
Above: This VF-1 Tomcat was photographed at NAS Miramar, September 1974, prior to the squadron's first cruise with the Tomcat. It has been claimed that during this cruise, the *Wolfpack* strafed ground targets during *Operation Frequent Wind*, the evacuation of Saigon. This author could find nothing to substantiate this in the squadron's command history. This F-14A later flew with VF-124, VF-2, PMTC, and at NADC Warminster. It was stricken 7 April 1992. (Author's Collection)

F-14A-70-GR, **158999**
Right: The 60th Tomcat was photographed December 1973, at Grumman's Calverton Facility, wearing only a primer finish. The only markings consist of white control surfaces, NAVY and the number 60. Of interest is the Northrop Television Camera System. This F-14A was assigned to NAS Miramar where it served with VF-2, VF-124, VF-1, and VF-301. On 29 November 1993 this Tomcat logged VF-301's 70,000 mishap-free flight hour during an ACM sortie. It was transferred to VF-202 in 1993 and stricken 6 September 1994. This Tomcat is preserved at NAS Fort Worth. (McNeil via Paul)

F-14A-70-GR, **159000**
Right: The Red Lightnings of VF-194, had a very short association with the Tomcat. The squadron stood up at NAS Miramar on 1 December 1986. In April 1988, falling victim to budget cuts, it was disestablished before making its first operational cruise. This example served with VF-1 and VF-124 before being assigned to VF-194. It reverted back to VF-124 and was retired to AMARC on 20 August 1990. It was stricken 17 October 1994. (Snyder)

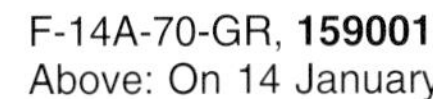
F-14A-70-GR, **159001**
Above: On 14 January, 1975 *Wolfpack* Pilot LCDR David Bjerke and NFO LT Gerald Kowlok launched from the USS Enterprise (CVN-65) for a VFR air intercept mission. An hour into the flight the crew experienced vibration and observed smoke and flames accompanied by uncontrollable yaw. The crew ejected and were rescued. The crash was the result of another failure in the TF30-P-412A's compressor section. As a result, all Tomcats were briefly grounded pending an investigation of the troublesome powerplant. (Author's Collection)

F-14A-70-GR, **159002**
Left: Established at NAS Miramar, on 1 December 1986, *Satan's Kittens* of VF-191 were assigned to the same carrier air wing, CVW-10 as VF-194. Both squadrons went away 30 April 1988. This Tomcat, like 159000, also served with VF-2 and VF-124. It reverted back to VF-124 before going to AMARC on 17 August 1990. It was stricken 17 October 1994. (Trombecky/Airframe Images)

F-14A-70-GR, **159003**
Left: This image of a VF-302 Tomcat demonstrates how weathering and operational use affect the low-viz paint scheme. This Tomcat spent its entire service life on or near the west coast. It was assigned to VF-124, VF-1, VF-301, and VF-302, the squadron depicted in this image. This Tomcat was retired to AMARC on 10 September 1990. It was stricken 17 October 1994. (Rys)

F-14A-70-GR, **159004**
Right: Delivered 27 April 1974, this F-14A was initially assigned to Pacific Fleet squadrons, VF-124 and VF-1. In 1982 it joined the *Diamondbacks*, of VF-102, an Atlantic Fleet squadron. It was transferred to VF-101 in 1990. Following service with VF-41, it was delivered to AMARC on 27 September 1991. It was stricken 17 October 1994 and photographed at AMARC in April 1996. (Author)

F-14A-70-GR, **159005**
Right: Tomcats had already served nearly ten years in the fleet prior to reaching Naval Reserve squadrons like VF-301. This Tomcat was photographed in August 1985 at NAS Fallon Nevada where it was taking part in a squadron strength training deployment. Reserve squadrons trained just as hard as the regular fleet squadrons and were no less capable. This example also saw service with VF-124, VF-1, and VF-24. It was retired to AMARC 24 September 1991 and stricken 17 October 1994. (Grove)

F-14A-70-GR, **159006**
Below: This was the final block 70 Tomcat constructed. All with the exception of 158984 have been stricken or lost through attrition. This example was delivered to VF-124 on 18 April 1974. It flew with VF-2 prior to joining VF-102. It went to AMARC in November 1991 and was stricken 17 October 1994. (Author's Collection)

F-14A-75-GR, **159007**
Left: This VF-124 Tomcat was photographed on 28 May 1974, shortly after its delivery to the squadron. It was utilized to train aircrews of VF-14, then in the process of transitioning from the F-4B. It was delivered to NAS Oceana in July 1974. On 5 August 1975, while operating from the USS John F. Kennedy (CV-67), it overran the deck following arrester gear failure and was lost overboard. (Curry)

F-14A-75-GR, **159008**
Below: According to VF-32's official history, this F-14A was the first Tomcat assigned to an Atlantic Fleet Squadron. It was initially delivered to VF-124 on 18 March 1974 and utilized to train VF-32 crews while they transitioned from the F-4B. On 14 June 1974, it was transferred to VF-32. Later assigned to VF-101, it was stricken on 4 November 1977 when it crashed near Oceana. (Author's Collection)

F-14A-75-GR, **159009**
Left: During June 1975, the *Swordsmen* of VF-32 made the first Atlantic Fleet F-14 deployment aboard the USS John F. Kennedy (CV-67). During that cruise, VF-32 received the Admiral Joseph Clifton Award honoring it as the Navy's top squadron. When photographed on 6 April 1976, this Tomcat wore a mix of Bicentennial markings and a number one signifying receipt of the Clifton award. It later served with VF-41, VF-31, and VF-101. It went to AMARC on 31 July 1991 and was stricken 17 October 1994. (JEM Slides)

F-14A-75-GR, **159010**
Right: The gloss gray paint scheme on this VF-11 Tomcat has become a bit mottled due to corrosion control efforts. The "E" and "S" on the rudder signify the COMNAVAIRLANT Battle "E" and CNO Safety "S." Originally delivered to VF-124 in April 1974, this Tomcat was used to train VF-32 crews who brought it to NAS Oceana in July 1974. It has also served with VF-142, VF-101, and VF-33. It retired to AMARC in 31 May 1991 and was stricken 17 October 1994. (JEM Slides)

F-14A-75-GR, **159011**
Right: This VF-11 Tomcat was stricken 6 February 1982 when it crashed into the Red Sea during an ACM mission while taking part in the squadron's first cruise aboard the USS John F. Kennedy (CV-67). Delivered to VF-124 on 25 April 1974, this Tomcat later served with VF-14 and VF-101 before joining VF-11 during December 1980. It was photographed at NAS Oceana on 30 April 1979 while assigned to VF-14. (Tunney)

F-14A-75-GR, **159012**
Below: Photographed on 29 May 1974, this Tomcat was being utilized to train VF-14 aircrews. Along with its sister squadron VF-32, these were the first two Atlantic Fleet Tomcat squadrons. The Tomcat was flown to NAS Oceana and assigned to VF-14 on 25 July 1974. It later served VF-101 and ended its flying career with the *Tophatters* of VF-14. It was stricken 20 March 1978, when it crashed into the Atlantic off the coast of Florida. (Curry)

F-14A-75-GR, **159013**
Left: This Tomcat, callsign *Gypsy* 204, was photographed on 29 January 1989, returning from an historic MED cruise aboard the USS John F. Kennedy (CV-67). During this deployment, the *Swordsmen* of VF-32 downed a pair of Libyan MiG-23 Flogger Es. A little more than a year after this historic event, this F-14A was retired to AMARC. It was stricken 17 October 1994. (Author)

F-14A-75-GR, **159014**
Left: This F-14A was delivered to VF-124 on 29 May 1974 and utilized to instruct VF-14 crews then making the transition to the Tomcat. It later served with VF-101 and VF-11. It was retired to AMARC 23 July 1991, following service once again with VF-14. It was photographed at AMARC on 27 April 1998. (Author)

F-14A-75-GR, **159015**
Below: On 13 November 1974, this VF-32, Tomcat, crewed by Cdr. J.G. Knutson and Ltjg. D.C. Leestma, made the first F-14A arrested landing aboard the USS John F. Kennedy (CV-67). It later served with VF-101, VF-33, and VF-11 before reaching AMARC on 4 September 1991. It was stricken 17 October 1994. (Author's Collection)

F-14A-75-GR, **159016**
Right: In 1982, the *Swordsmen* redesigned their tail markings. The new design, the brainchild of VF-32 Corrosion Control, featured a Gypsy Tomcat. And of course, like the Tomcat, it had twin tails. New markings such as these were a welcome change. This F-14A was photographed in April 1983 following its first cruise aboard USS Independence (CV-62). It was retired to AMARC on 15 July 1991 and stricken on 17 October 1994. (Author)

F-14A-75-GR, **159017**
Above: Delivered to VF-124 on 12 June 1974, this Tomcat later served with VF-14, VF-101, VF-11, and VF-102. It was photographed at NAS Oceana in October 1981 while assigned to the *Red Rippers* of VF-11. It arrived at AMARC on 24 September 1991 and was stricken 17 October 1994. (Author)

F-14A-75-GR, **159018**
Right: This unmarked Tomcat displays the standard paint scheme applied to Tomcats delivered in 1974. This example was photographed at Sheppard AFB, Texas, 27 October 1974. Most likely it was being flown by a VF-32 crew from NAS Miramar on route to NAS Oceana. While assigned to the Atlantic Fleet this Tomcat was operated by VF-32, VF-101, VF-143, VF-11, VF-14, and VF-142. It arrived at AMARC 30 April 1991 and was stricken 17 October 1994. (Mills)

F-14A-75-GR, **159019**
Above: Even the subdued light of a hangar cannot detract from the overall appearance of this high-vis paint scheme applied to this VF-14 Tomcat. Delivered to VF-14 on 21 June 1974, it later served with VF-31, VF-101, VF-102, and VF-32 before being transferred to the Naval Reserve and VF-301. It was stricken 14 April 1994. (Author's Collection)

F-14A-75-GR, **159020**
Left: Delivered to VF-124 on 17 July 1974, this F-14A was used to instruct crews from VF-14, the Navy's oldest active squadron. On 4 November 1974, the *Tophatters* flew this Tomcat to NAS Oceana. While assigned to the Atlantic Fleet it served with VF-41, VF-31, VF-101, VF-33, and VF-74. It was flown to AMARC on 12 August 1991 and stricken 17 October 1994. It was photographed at that venue, wearing VF-74 markings and storage code AN1K0037 in April 1996. (Author)

F-14A-75-GR, **159021**
Left: AC/105, an F-14A assigned to the *Red Rippers* of VF-11, taxies out for another training sortie during air wing work-ups for a January 1982 cruise aboard the USS John F. Kennedy (CV-67). This example was delivered to VF-124 on 3 July 1974. It went to VF-32 in October the same year. Since arriving at NAS Oceana, this Tomcat has flown with VF-101, VF-11, VF-33, and VF-143. It was dispatched to AMARC on 26 September 1991 and was stricken 17 October 1994. (Grove)

F-14A-75-GR, **159022**
Above: This overall gloss gull gray Tomcat was assigned to VF-32 when photographed at NAS Oceana 30 April 1979. The "AB" tail markings indicate it is assigned to CVW-1, aboard the USS John F. Kennedy (CV-67). On 5 December 1979 while assigned to VF-14, it collided with F-14A 160902 and was lost near Roosevelt Roads, Puerto Rico. (Tunney)

F-14A-75-GR, **159023**
Right: This VF-14 Tomcat was photographed in May 1983, five months prior to the *Tophatters* involvement in *Operation Urgent Fury,* the liberation of Grenada. From 23 October to 5 November 1983, the *Tophatters* flew 82 combat missions. These mainly consisted of Combat Air Patrol and escort for their sister squadron, VF-32, which flew TARPS missions. This Tomcat spent its entire operational career at NAS Oceana assigned to VF-14, VF-101, and VF-84. It completed its flying while assigned to the NATC and was stricken 2 April 1992. (Author)

F-14A-75-GR, **159024**
Right: This VF-14 Tomcat was taking part in an air show at FAA's Pomona New Jersey facility when photographed 20 August 1977. It was later transferred to VF-31 during April 1981. It was stricken 8 November 1983, when it crashed 75 miles south of Cyprus while on a low-level flight from the USS John F. Kennedy (CV-67). (Spering)

F-14A-75-GR, **159025**
Above: This Tomcat was delivered to VF-124 on 2 August 1974. Reassigned to VF-32, it was flown to NAS Oceana 13 November 1974. It has since served with VF-101 and VF-11. It was photographed in VF-32 markings on 24 March 1979. Flown west in June 1984 it joined VF-302, a Naval Reserve squadron, and it finished its flying days with VF-202. Stricken 6 September 1994, it is currently listed as preserved in Charleston, S.C. (Huston)

F-14A-75-GR, **159421**
Left: This engineless VF-14 CAG bird was photographed at NAS Oceana in April 1976 wearing this eye-catching Bicentennial paint scheme. The *Tophatters* took delivery of this Tomcat on 10 August 1974. It later served with VF-101 and VF-31. It was stricken 18 June 1987 while flying with VF-101. (JEM Slides)

F-14A-75-GR, **159422**
Left: Even bright sunlight cannot add contrast to this low-vis paint scheme. It must be effective. This VF-302 Tomcat was photographed in June 1991. The *Stallions* assumed the adversary mission in the 1990s, becoming renowned for their prowess in ACM. VF-302 was disestablished 11 September 1994. Prior to its assignment to the Naval Reserve, this Tomcat served with VF-14, VF-143, and VF-31. Delivered to AMARC on 26 July 1991, it was stricken 17 October 1994. (Grove)

F-14A-75-GR, **159423**
Right: Atlantic Fleet Tomcat training was in high gear when this VF-101 Tomcat was photographed in October 1978. Of interest is the tailcode which is shadowed in gold. Following delivery to VF-124 on 20 September 1974, this example was flown to NAS Oceana by VF-32 on 10 October 1974. It returned to NAS Miramar in 1989 and was assigned to VF-302. Entering AMARC on 6 December 1990, it was stricken on 17 October 1994. (Brown)

F-14A-75-GR, **159424**
Right: Not all tired Tomcats made it to AMARC. This example, photographed at PMTC during October 1995, was in the process of being cannibalized for useful parts. It had been stricken 25 February 1994. This example was delivered to NATC on 25 August 1974 and spent its entire service life performing the RDT&E mission, first for NATC and then with VX-4. (Author)

F-14A-75-GR, **159425**
Below: Wearing a very faded gull gray over white paint scheme, this VF-14 Tomcat was photographed during October 1977. The *Tophatters* received this example from Grumman 7 October 1974. It was later assigned to VF-101, VF-33, and VF-74. Sent to rest at AMARC in June 1991 it was stricken 17 October 1994. (Dorr)

F-14A-75-GR, **159426**
Left: Delivered to VF-124 from Grumman on 18 September 1974, this Tomcat was transferred to VF-32 on 10 October 1974. It would later serve with VF-101 and VF-33 prior to going to NADEP Norfolk in 1986. It flew test sorties with PMTC, from September 1987 until entering AMARC on 29 November 1990. It was stricken 17 October 1994. (Jim Wooley collection)

F-14A-75-GR, **159427**
Left: Originally known as the *Tarsiers,* tail markings on this VF-33 Tomcat denote the squadron's new nickname – *Starfighters.* The squadron received its first Tomcat on 20 January 1982. This example was photographed in March 1985 while working up for a cruise aboard USS America (CV-66). This Tomcat had been assigned to VF-124, VF-14, and VF-101 prior to reaching VF-33. Delivered to AMARC on 13 September 1991 it was stricken on 17 October 1994. (Grove)

F-14A-75-GR, **159428**
Below: This Tomcat was assigned to VF-33 when it was photographed at NAS Oceana in May 1982. This attractive paint scheme was the result of a contest staged by VF-33. This Tomcat was damaged on 17 June 1984 when its undercarriage collapsed while landing on the USS America (CV-66). Delivered to VF-124 on 8 October 1974, it was transferred to VF-101 on 8 December 1975. It later served with VF-14, VF-33, VF-202, and VF-201. As of February 1998 it was assigned to VF-101. (Author)

F-14A-75-GR, **159429**
Right: A very small NADC logo on the tail of this F-14A denotes its assignment to the Naval Air Development Center, Warminster, Pennsylvania, when it was photographed 22 September 1991. Delivered to VF-124 on 8 October 1974, it went on to fly with VF-101, VF-32, and VF-33. Delivered to AMARC on 16 January 1992 it was stricken on 17 October 1994. (Roop)

F-14A-80-GR, **159430**
Below: The *Pukin' Dogs* of VF-143 picked up their unusual sobriquet during the Vietnam War when a F-105 driver coined the term when referring to the squadron's winged lion. Originally based on the west coast, the squadron moved to NAS Oceana in 1974, the same year it made the transition from the F-4 Phantom to the Tomcat. This example was photographed in August 1975, at NAS Oceana. It crashed while flying from the USS Eisenhower (CVN-69) on 5 October 1978. (Spering)

F-14A-80-GR, **159431**
Right: Colorful squadron markings were the norm, not the exception in 1977, when this VF-142 Tomcat was photographed aboard the USS America (CV-66). Following delivery to VF-124 on 21 October 1974, it was assigned to VF-142 on 3 June 1975. It later flew with VF-101 and VF-14. It was stricken on 3 January 1987 during VF-14's MED Cruise aboard the USS John F. Kennedy (CV-67). (Shapcott via Sheets)

F-14A-80-GR, **159432**
Left: Delivered to NAS Miramar 13 November 1974, this Tomcat was utilized to train VF-143 pilots, NFOs and maintainers and was transferred to that squadron on 1 May 1975. This Tomcat was photographed at NAS Glenview early in June 1975. On 26 June 1975 it suffered a first stage turbine failure resulting in an explosion and fire while taking off from NAS Oceana. Extensively damaged, the shell was shipped to NADEP Norfolk, 21 July 1975, and it was stricken 28 August 1976. Designated the Structural Life Test Article, it was used to determine the ultimate structural life of the Tomcat, and what modifications were necessary to achieve it. As such, this Tomcat became one of the most important F-14s in the fleet. (Moore via LeBaron)

F-14A-80-GR, **159433**
Above: This Tomcat began and ended its operational career with the *Ghostriders* of VF-142. Delivered to the squadron at NAS Miramar on 4 April 1975, it was the first Tomcat received by the squadron. It also served with VF-101 and VF-14 before returning to VF-142 in September 1989. Delivered to AMARC on 15 November 1991, it was stricken on 17 October 1994. (Curry)

F-14A-80-GR, **159434**
Left: Tomcats have been thrilling air show crowds since being introduced to squadron service. The *Grim Reapers* of VF-101 usually provided the aircraft and crews for east coast air show performances. This example, wearing a less than thrilling paint scheme, flew at the 1988 McGuire AFB, open house. It was delivered 9 November 1974 as the 100th Tomcat built. It served with VF-124, VF-143, and VF-11. It arrived at AMARC on 18 December 1991 and was stricken 17 October 1994. (Author)

F-14A-80-GR, **159435**
Right: Delivered to VF-124, on 28 October 1974, this Tomcat was photographed at NAS Oceana in August 1975, shortly after VF-143 was transferred to the Atlantic Fleet. It later served with VF-14 and VF-101 and was delivered to AMARC on 31 October 1991. It was stricken 17 October 1994. (Author's Collection)

F-14A-80-GR, **159436**
Right: Delivered to VF-124 on 16 November 1974, this VF-142 Tomcat spent the remainder of its service life assigned to east coast squadrons, VF-101, VF-14 VF-142, and VF-143. It was photographed aboard the USS America (CV-66) during a 1977 Atlantic cruise. Following a brief period of storage at NADEP it was delivered to AMARC on 19 November 1991. It was stricken 17 October 1994. (Shapcott via Sheets)

F-14A-80-GR, **159437**
Below: Although VF-142 did not make its first carrier cruise until 15 April 1976, this *Ghostrider's* Tomcat displays "USS America" on its tail in August 1975. It remained at NAS Oceana, serving with VF-31, VF-32, and VF-101. During service with VF-32 it took part in the shoot-down of a pair of Libyan MiG-23 Floggers. Delivered to AMARC on 11 February 1992, it was stricken like so many other F-14As, with the stroke of a computer key, on 17 October 1994. (Author's Collection)

F-14A-80-GR, **159438**
Above: Assigned to CVW-6, with VF-142, the *Pukin' Dogs* of VF-143 made their first Tomcat cruise aboard the USS America (CV-66), 15 April 1976, to 25 October 1976. Following delivery to VF-124 on 12 December 1974, this Tomcat spent its entire career at NAS Oceana, flying with VF-11, VF-101, and VF-102. It was stored at AMARC starting 18 March 1992 and was stricken 17 October 1994. (Author's Collection)

F-14A-80-GR, **159439**
Left: Minus both engines, this VF-142 Tomcat rests in a maintenance hangar at NAS Oceana, August 1975. Due to a number of operational losses, the Tomcat's TF-30 turbofans were under constant scrutiny. The TF-30s were intended as an interim engine and were not ideally matched to the F-14's highly advanced aerodynamic design. To date, more than 144 Tomcats have been written off, more than 25% due to problems encountered with the early TF-30 turbofans. Following service with the *Ghostriders*, this Tomcat was transferred to VF-101. It crashed 15 July 1984 while assigned to VF-102. The crew experienced an inflight emergency and was forced to eject. (Author's Collection)

F-14A-80-GR, **159440**
Left: This Tomcat was delivered to VF-124 on 7 December 1974, and aside from a brief stint with PMTC in 1986, it spent its career flying out of NAS Oceana, assigned to VF-142, VF-14, VF-102, VF-101, VF-33 before returning to VF-142. in 1987. Photographed in VF-142 markings during July 1976, this Tomcat arrived at AMARC on 5 March 1992 and was stricken 17 October 1994. (Leslie via Rotramel)

F-14A-80-GR, **159441**
Above: In order to contrast paint schemes, compare this low-vis scheme to that of VF-32 Tomcat 159016. Both were photographed in May 1983. This example was delivered to VF-124 on 4 December 1974 and served with VF-142 and VF-31. This Tomcat crashed on 25 March 1988 when it failed to develop sufficient airspeed after being launched from the USS John F. Kennedy (CV-67). (Author)

F-14A-80-GR, **159442**
Right: Attempts at corrosion control have given this VF-143 Tomcat a very weathered appearance. At least you can still make out the bureau number. This Tomcat was assigned to VF-101 and VF-301 prior to being interred at AMARC on 20 September 1990. It was stricken 17 October 1994. (Roop)

F-14A-80-GR, **159443**
Right: Delivered to VF-124 on 17 December 1974, this Tomcat was transferred to VF-143. It crashed into the Atlantic following a ramp strike aboard the USS America (CV-66), on 28 March 1977. (Van Aken)

F-14A-80-GR, **159444**
Above: The red flashing lights below the radome and on the tail indicate this VF-143 Tomcat is powered up by a ground unit. Any system as complex as the F-14 Tomcat requires many hours of maintenance for every flight hour. The "Tomcat Tweakers" pictured here are the unsung heroes who keep the Tomcats airworthy. This example also served with VF-101, VF-32, and VF-14 before retiring to AMARC on 1 April 1992. It was stricken 17 October 1994. (Spering)

F-14A-80-GR, **159445**
Left: On 17 April 1991, CVW-1 returned from the Persian Gulf and *Operation Desert Storm*. The *Starfighters* of VF-33 deployed 28 December 1990, six months ahead of schedule, arriving 16 January 1991, the day before the air war kicked off. The squadron's first mission on 19 January consisted of a MiG sweep and HVU CAP for a strike against the Latifiya Scud missile production facility, which was destroyed. Each participating VF-33 aircrew was awarded the Navy Commendation Medal with combat distinguishing device. This Tomcat was transferred to VF-101 and was stricken 8 September 1994. (Author)

F-14A-80-GR, **159446**
Left: The "roof crew" prepares this VF-142 CAG bird for launch aboard the USS America(CV-66), in August 1976. Of interest are the Phoenix, Sparrow and Sidewinder missiles. This was the *Ghostriders* first cruise since transitioning to the Tomcat. This example was later assigned to VF-101 and VF-33. It was stricken on 3 April 1992. (Jay)

F-14A-80-GR, **159447**
Right: A VF-102 F-14A flies formation with a USAF KC-135A during January 1986. Of interest is the drogue modification to the Stratotanker's flying boom which permits refueling of Navy aircraft. The *Diamondbacks* were the last squadron to operate this Tomcat. It arrived at AMARC on 17 April 1992 and was stricken 17 October 1994. (USN)

F-14A-80-GR, **159448**
Below: When photographed in October 1990, VF-33 was preparing for its second deployment in less than a year. The *Starfighters* departed for the Persian Gulf, 28 December 1990, arriving 16 January 1991, one day before *Desert Shield* became *Desert Storm*. This example was transferred to NADC Warminster, and was stricken 14 June 1993. (Sagnor)

F-14A-80-GR, **159449**
Right: The era of high-vis paint schemes witnessed some beautifully marked aircraft, such as this VF-142 Tomcat. It was photographed prior to the crowds arriving for the 1978 NAS Norfolk open house. This example also served with VF-101, VF-31, and VF-14. It arrived at AMARC on 13 May 1992. It was stricken 17 October 1994. (Jay)

F-14A-80-GR, **159450**
Left: By 1979, *Pukin' Dogs* markings had been toned down considerably. Compare these VF-143 markings with 159435 photographed two years earlier. Unfortunately, for Tomcat enthusiasts, colorful markings would continue to disappear, becoming almost nonexistent. Delivered 8 February 1975, this Tomcat served with VF-124, VF-143, VF-101, and VF-33. It was stricken on 31 May 1994. (McIntosh)

F-14A-80-GR, **159451**
Left: Delivered to VF-124 on 7 March 1975, this Tomcat was assigned to VF-142 at NAS Oceana two months later. It was stricken 11 November 1977 following a collision with VAQ-137 EA-6B Prowler 158809, near the USS America (CV-66) operating in the Mediterranean Sea. (Greby Collection)

F-14A-80-GR, **159452**
Below: From 10 March until 10 September 1986, the USS John F. Kennedy (CV-67), participated in a MED Cruise. The *Diamondbacks* of VF-102 formed half of the fighter contingent of the embarked air wing, CVW-1. This example was photographed during a June 1986 Italian port of call. It was stricken on 21 February 1995. (via Henderson)

F-14A-80-GR, **159453**
Right: Personal markings such as the rattlesnake applied to this VF-142 Tomcat were rare. Its significance is unknown. It may be another squadron's zap, possibly applied by the *Diamondbacks* of VF-102. While based at NAS Oceana, this Tomcat also served with VF-101 and VF-14. It arrived at AMARC on 29 May 1992 and was stricken 17 October 1994. (Rys)

F-14A-80-GR, **159454**
Right: The TACTS and ACMI pods represent the only color on this low-vis VF-201 Tomcat. The *Hunters* were established 25 July 1970, at NAS Dallas. The squadron began its transition to the Tomcat in 1987. With the *Superheats* of VF-202, the *Hunters* made up the fighter element of CVWR-20. This example was delivered to VF-124 on 15 March 1975. It served with VF-143, VF-101, VF-11, VF-302, and VF-301 and entered storage at AMARC on 23 July 1994. It was stricken 17 October 1994. (Grove)

F-14A-80-GR, **159455**
Below: This example was delivered to VF-124 on 10 March 1975. It later flew with VF-1, VF-24, and VF-143. It entered NADEP Norfolk in 1983 and was delivered to NATC during March 1987. From 14 November to 20 December 1990, this NF-14A was utilized to explore the air-to-ground capabilities of the Bombcat. During that time frame a total of 23 flights were made hauling a variety of air-to-ground ordnance. This Tomcat has since been stricken on site at Patuxent River. (Pugh)

F-14A-80-GR, **159456**
Left: Delivered to VF-124 on 25 February 1975, this F-14A was utilized to train personnel of VF-143. It arrived at NAS Oceana during May 1975 assigned to the *Pukin' Dogs*. This Tomcat was photographed 11 May 1978 while assigned to VF-101. It was stricken 21 May 1979 when it crashed during a MED Cruise while embarked aboard the USS Eisenhower (CVN-69). (Dorr)

F-14A-80-GR, **159457**
Left: This example was delivered to VF-124 on 29 March 1975. It later served with VF-143, VF-14, VF-32, and VF-101. It is pictured here in the markings of VF-14 on 19 September 1987. It was stricken 23 September 1993. As of September 1997 it was a stripped-out hulk parked on the north side of NAS Oceana. (Van Aken)

F-14A-80-GR, **159458**
Below: This VF-102 F-14A wore vestiges of color when it was photographed aboard the USS America (CV-66), during a 1986 MED Cruise. This example was transferred to VF-101 and stricken 25 August 1994. As of September 1997 it was a stripped-out hulk parked on the north side of NAS Oceana. (Author's Collection)

F-14A-80-GR, **159459**
Right: Proving that Carrier Air Wings carry out flight operations around the clock, this outstanding image of a VF-102 Tomcat in zone 5 afterburner was taken by noted author Robert F. Dorr, aboard the USS America, 31 August 1989. This F-14A was delivered to VF-124 on 21 April 1975. It later flew with VF-1, VF-2, and VF-24 before its assignment to VF-102. It finished its flying career with VF-101 and was stricken 7 April 1992. (Dorr)

F-14A-80-GR, **159460**
Above: The *Diamondbacks* of VF-102 transitioned to the F-14A during 1981. This example arrived at Oceana during April 1982 following service with VF-124, VF-2, and VF-211. It was photographed in *Diamondbacks* markings in October 1984. Following service with VF-124 this Tomcat was delivered to AMARC 20 November 1990. It was stricken 17 October 1994. (Author)

F-14A-80-GR, **159461**
Right: *Wolfpack* groundcrew go through pre-flight checks prior to launching this F-14A during November 1975. This aircraft was stricken 24 March 1976. While on a local training mission near NAS Miramar the crew, pilot LCdr. Clareance Irvin and RIO Lt. Steve Sabin, ejected following a loud thump followed by loss of control. Following this crash, all F-14As were grounded. As a result of an investigation, flight restrictions were placed on the Tomcat's TF-30-P412A powerplants. (Author's Collection)

F-14A-80-GR, **159462**
Left: This factory-fresh VF-124 Tomcat was photographed during July 1975, three months after being delivered from Grumman. It was utilized to train VF-2 personnel and later went to that squadron. It flew with VF-211 before transferring to the Atlantic Fleet and VF-32. Its final squadron was VF-143. They delivered it to AMARC on 5 August 1992. It was stricken 17 October 1994. (McGarry)

F-14A-80-GR, **159463**
Above: This image depicts the markings common to most Tomcats operated by VF-124 during the mid 70s. The *Gunfighters* were dedicated to training naval aviators for nearly forty years. From 1958 until 1972, the squadron trained prospective F-8 Crusader pilots and maintainers. Their first F-14 Tomcat arrived in October 1972. This example also served with VF-1 and VF-24 before joining VF-101, the Atlantic Fleet Readiness Squadron. The last operational squadron to fly it was VF-14. It was stricken 13 August 1992. (Author's Collection)

F-14A-80-GR, **159464**
Left: Delivered to VF-124 on 4 April 1975, this aircraft was stricken the following year while assigned to VF-2. Wearing the code NK/213, it crashed while landing aboard USS Enterprise (CVN-65), steaming near the Philippines on 19 December 1976. (Trombecky/Airframe Images)

F-14A-80-GR, **159465**
Above: This F-14A was delivered to VF-124 on 7 May 1975. Within three months it had been reassigned to VF-1. It later served with Atlantic Fleet squadron VF-102. In 1991 it returned to NAS Miramar to serve with VF-302. As of February 1998 it was once again training Tomcat aircrews with the *Grim Reapers* of VF-101. (Author's Collection)

F-14A-80-GR, **159466**
Right: This overall gloss gull gray VF-102 CAG bird was photographed at NAS Oceana during May 1982. It previously served with VF-124, VF-2, and VF-211. On 1 October 1992 it was delivered to AMARC. It was stricken 17 October 1994. (Author)

F-14A-80-GR, **159467**
Right: This *Wolfpack* Tomcat was photographed during August 1977 on the Buckley ANGB transient line. It was assigned to CVW-14 aboard the USS Enterprise (CVN-65). That carrier's name appears just aft of the glove vane. This Tomcat entered AMARC on 5 October 1994. Before heading to its desert resting place it served with VF-124, VF-1, VF-211, VF-2, VF-101, VF-143, and finally, the *Jolly Rogers* of VF-84. (Stumpf via Sheets).

F-14A-80-GR, **159468**
Left: Loaded with a pair of external fuel tanks, this VF-143 Tomcat was photographed aboard the USS Eisenhower (CVN-69) during CVW-7's 1988 MED cruise. This Tomcat was the last block 80 constructed. It also served VF-124, VF-2, VF-24, and VF-101. It arrived at AMARC 29 June 1994 and was stricken 13 December 1994. (Bottaro via Snyder)

F-14A-85-GR, **159588**
Above: This Tomcat was stricken on 14 September 1976, while assigned to VF-32, operating off the Orkney Islands, near the northeast coast of Scotland. Due to a problem with its auto throttle, it taxied off the deck of the USS John F. Kennedy (CV-67). The Tomcat armed with AIM-54 Phoenix missiles plunged into the deep waters of the North Atlantic. Following extensive search and salvage operations, the Tomcat and its missiles were recovered. The nets tangled around the fuselage were from previous attempts to snare this Tomcat. (USN)

F-14A-85-GR, **159589**
Left center: Returning from *Operation Desert Storm*, Diamondback/106 taxies at NAS Oceana on 17 April 1991. This was not the first combat deployment for VF-102 Tomcats. During *Operation Prairie Fire* VF-102 Tomcats were targeted by Libyan anti-aircraft artillery and SA-5 missiles. Later, on 15 April 1986, the squadron flew top cover for *Operation El Dorado Canyon*. This Tomcat went to AMARC on 3 December 1991. It was later stricken on 17 October 1994. Besides VF-102 it served with VF-124, VF-32, VF-101, VF-143, and VF-33. (Author)

F-14A-85-GR, **159590**
Left: Delivered on 31 May 1975, this F-14A was the first VF-124 Tomcat to suffer a class A mishap. It was stricken on 29 October 1975, while operating from the USS Enterprise (CVN-65). The Tomcat reportedly caught fire when the third stage fan of the port engine suffered a catastrophic failure. The crew ejected successfully. (Trombecky/Airframe Images)

F-14A-85-GR, **159591** (TARPS)
Right: The 1988 Reconnaissance Air Meet was held at Bergstrom AFB during the first two weeks of August. The Navy participated with TARPS equipped Tomcats from VF-202, VF-302, and VF-101. The *Stallions* of VF-302 received awards for the Top Navy Squadron and the Best Navy Crew, which consisted of Cdr. Mel Logan and Lt. Dave Fuller. Prior to reaching VF-302, this Tomcat served with VF-124, VF-101, VF-142, and VF-32. As of February 1998 it was assigned to VF-201 at NAS Fort Worth JRB. (Author)

F-14A-85-GR, **159592** (DR-10)
Right: The *Fighting Renegades* of VF-24 received their first Tomcat in November 1975. This example, photographed in June 1981, carries a CATM-7, the training version of the AIM-7 Sparrow. It was assigned to VF-24 first in 1976 and again in 1981. It also served with VX-4 and VF-21. In January 1990 it returned to Grumman and later rejoined the Pacific Fleet in March 1993 as the 10th F-14D(R). The (R) indicates the airframe has been remanufactured. As of February 1998 this Tomcat was serving with VF-101. (Grove)

F-14A-85-GR, **159593**
Below: This VF-24 Tomcat has managed to retain sufficient color to make it distinctive. The nine on the stabilizer indicates the squadron is assigned to CVW-9, aboard the USS Constellation (CV-64). On 20 May 1985, this Tomcat became the first to log 3000 flight hours. It would log more hours with VF-101, VF-14, and the Fighter Weapons School before being stricken 6 October 1993. (GB Slides)

F-14A-85-GR, **159594**
Left and below: Records indicate this Tomcat was the first example delivered to VF-101 when it was transferred from VF-124 on 1 April 1976. It crashed into the Atlantic Ocean 21 June 1977 following an engine failure 48 miles southeast of NAS Oceana. It was photographed 30 June 1977 lashed to the side of the salvage ship USS Recovery, (ARS-43). (LeBaron/USN Photo)

F-14A-85-GR, **159595** (DR-12)
Left: This example, photographed in VF-11 markings, formerly served with VF-124, VF-101, VF-143, VF-14, and VF-142 before entering NADEP, Norfolk 30 March 1990. It emerged in 1994, as the 12th F-14D(R). Compare this example with 159592. The most striking difference is the afterburner cans of the General Electric F110-GE-400 turbofans and the Martin-Baker MB-14 ejection seats. As of February 1998 it was flying with VF-2. (Grove)

F-14A-85-GR, **159596**
Left: Before arriving at AMARC on 12 December 1990 this F-14A served with VF-124, VF-101, VF-142, VF-14, and the NATC. It was photographed at NAS Oceana in early VF-142 markings. This example was stricken 17 October 1994. (USN Photo)

F-14A-85-GR, **159597**
Right: The 144th Tomcat sits on the ramp at Grumman's Calverton Facility waiting for a crew from VF-124 to fly it to NAS Miramar on 26 July 1975. It was utilized to train VF-14 aircrews and maintainers before joining that squadron on 13 August 1976. It served VF-101 from 1979 to 1983 and returned to the East Coast FRS again in 1989. Flown to AMARC on 8 December 1994, it was stricken there 17 July 1995. (Paul)

F-14A-85-GR, **159598**
Right: This very toned down VF-32 Tomcat was photographed at NAF Washington in August 1985. The tail markings have gone through another transition. Gone is the *Gypsy* Tomcat and gloss gray paint scheme. Of interest is the chin mounted Television Camera System. VF-32 and its sister squadron, VF-14, were the first Atlantic Fleet squadrons to receive TCS. Following service with VF-32, this example was assigned to the PMTC before entering storage at AMARC on 29 November 1990. It was stricken 17 October 1994. (Author)

F-14A-85-GR, **159599**
Below: Delivered to VF-124 on 11 July 1975, this VF-32 Tomcat was later transferred to VF-101 and then VF-142. It was stricken 2 June 1982 while assigned to VF-143. The arrester gear cable failed as it was being recovered aboard the USS Eisenhower (CVN-69) and the aircraft was lost overboard. (Author's Collection)

F-14A-85-GR, **159600** (DR-5)
Left: By 1981, Grumman was delivering Tomcats in an overall gloss gull gray scheme. Corrosion control efforts give it a mottled appearance. This example, assigned to VF-142, also served with VF-124, VF-101, and VF-14. It entered NADEP Norfolk in 1990 and emerged in 1994 as the 5th F-14D(R). It was assigned to the *Bounty Hunters* of VF-2 where it has remained through February 1998. (Author)

F-14A-85-GR, **159601**
Above: This VF-32 Tomcat photographed at NAS Miramar in October 1977, carries the inscription "Friendly Seal Mountain" on its canopy rail. Its significance is not known to the author. The *Swordsmen* received this F-14A from VF-124 in 1976. It was transferred to VF-142 in 1979 and crashed into the Atlantic ocean while on approach to the USS Eisenhower (CVN-69) on 6 March 1980. (Huston)

F-14A-85-GR, **159602**
Left: Initially delivered to VF-124 on 26 July 1975, this Tomcat returned to the *Gunfighters* in 1987. Between those dates it served with VF-101, VF-211, and VF-114. Gone are the orange wingtip and stabilizer markings common to VF-124 Tomcats photographed during the 1970s. This F-14A was stricken 22 September 1993. (Van Aken)

F-14A-85-GR, **159603** (DR-14)
Above: This Tomcat was delivered to VF-124 on 30 July 1975. It was later transferred to VF-101 on 4 May 1976 and was photographed on 29 April 1979. It served with VF-142 and then VF-32 until July 1990 when it was sent to NADEP Norfolk for remanufacture. It emerged as the 14th F-14D(R) and was assigned to VF-11. By February 1998 it was again assigned to VF-101. (McIntosh)

F-14A-85-GR, **159604**
Right: This Tomcat was assigned to VF-201 when it was photographed in October 1991. At the time there were four Naval Reserve Squadrons flying the F-14A; VF-201, VF-202, VF-301, and VF-302. Budget cuts have reduced that number to one, the *Hunters* of VF-201. This Tomcat was delivered to NATC in August 1975. It would later serve with VF-124, VF-101, VF-143, VF-32, VF-33, and VF-201. It was retired to AMARC on 1 October 1992 and stricken 17 October 1994. (Grove)

F-14A-85-GR, **159605**
Right: Transferred from VF-124 on 15 April 1976, this was one of the first Tomcats received by VF-101. It was photographed at NAS Oceana in October 1977. On 13 September 1980, while assigned to VF-143, this Tomcat crashed during a ACM training flight when it departed controlled flight and entered a flat spin. (Author's Collection)

F-14A-85-GR, **159606** (TARPS)
Left: Parked near a row of VF-103 F-4 Phantoms, this VF-142 Tomcat was photographed at NAS Oceana, 11 May 1978. It was delivered to VF-124 on 10 August 1975. After being rewired for TARPS, it served VF-101, VF-142, and VF-302. This Tomcat was stricken on 28 October 1992. (Paul)

F-14A-85-GR, **159607**
Below: An Iranian Tomcat visiting *TOPGUN?* The Navy Fighter Weapons School, also known as *TOPGUN*, has a long history of applying interesting paint schemes to their aircraft. This example is painted in full Iranian markings and was photographed during May 1992. Delivered to VF-124 on 18 August 1975, it would later serve with VF-101, VF-32, VF-142, and VF-124. It was stricken 9 February 1994. (Grove)

F-14A-85-GR, **159608**
Left: The *Aardvarks* of VF-114 began to transition to the F-14A in December 1975. This example was photographed in 1988 at NAS Miramar. Markings indicate the squadron has been awarded the Battle "E" and Safety "S" as well as the Adm. Clifton award for being the top fighter squadron in the Navy. Delivered to VF-124 on 21 August 1975, this Tomcat went on to serve with VF-211, VF-124, and VF-101 before going to AMARC on 26 October 1994. It was stricken 21 March 1995. (Holmes)

F-14A-85-GR, **159609**
Right: *Starfighter*, AB/211 taxis in at NAS Fallon, Nevada, the "Biggest little Naval Air Station in the world." At this venue carrier air wings, in this case CVW-1, can workup for an operational cruise. This Tomcat was previously flown by VF-124, VF-14, and VF-101. It was stricken 24 September 1993. (Grove)

F-14A-85-GR, **159610** (DR-2)
Above: Photographed at NAS Oceana 31 January 1989, this VF-32 Tomcat is returning from a historic MED Cruise aboard USS John F. Kennedy (CV-67). Callsign *Gypsy* 207, it took part in the shoot down of a pair of Libyan MiG-23 Flogger Es on 4 January 1989. Of interest is the absence of kill markings anywhere on this *Swordsmen* Tomcat. This F-14A was later remanufactured by Grumman and emerged as an F-14D(R). As of February 1998 it was assigned to VF-31. (Author)

F-14A-85-GR, **159611**
Right: Delivered on 7 September 1975, this Tomcat was assigned to VF-124 for the majority of its service life. In 1976 and again in 1979 it served with VF-24. It served briefly with VX-4 prior to being stricken 20 June 1994. (Van Aken)

F-14A-85-GR, **159612** (TARPS)
Left: A rather small *Pukin' Dog* adorns the tail of this VF-143 Tomcat. Delivered to VF-101 in February 1976, this F-14A also served with VF-302. It was stricken 8 September 1994. (Stewart)

F-14A-85-GR, **159613** (DR-4)
Left: This rather vanilla F-14A, assigned to VF-24, was photographed on the transient ramp at NAF Washington, 11 October 1986. It later flew with VF-124 and VF-24 before transfer to Grumman, Bethpage for remanufacture as a F-14D(R). Initially issued to VF-11, it was reassigned to VF-2 in November 1994. It was still serving with the *Bounty Hunters* as of September 1997. (McGarry)

F-14A-85-GR, **159614**
Below: This F-14A was delivered to VF-124 on 23 September 1975. When photographed on 20 November 1976, it was assigned to the *Flying Checkmates* of VF-211. This Tomcat was reassigned to VF-124 in 1987. It arrived at AMARC on 20 December 1991 and was stricken 21 February 1995. (Logan)

F-14A-85-GR, **159615**
Above: VF-32 made its first cruise aboard the USS Eisenhower (CVN-69) in October 1994. But this F-14A was photographed in October 1977 with the "IKE's" name applied, while the squadron was assigned to the USS John F. Kennedy (CV-67). The explanation may lie with the 1st FW F-15A parked next to it. In October 1977, VF-32 became the first Tomcat squadron to fly dissimilar air combat missions against USAF Eagles. These ACM sorties were launched from the "IKE." Fittingly, this Tomcat finished its flying career flying DACM at the Fighter Weapons School. It was stricken 21 December 1994. (Author's Collection)

F-14A-85-GR, **159616**
Right: Most U.S. Navy squadrons celebrated the Nation's 200th birthday by painting at least one aircraft in a Bicentennial scheme. This eye-catching example, photographed June 1976, adorns a VF-124 Tomcat. Transferred to VF-14 on 21 August 1976, it later served with VF-32, VF-11, VF-101, and the NATC. It was stricken 12 September 1994. (Taylor via Dorr)

F-14A-85-GR, **159617**
Right: Landing at NAF Atsugi, Japan, this Tomcat was photographed 20 November 1977 following VF-24's first Tomcat cruise. Callsign *Nickel*/203 was stricken on 14 June 1982 when it crashed into the Pacific Ocean near San Clemente Island. (Kudo via Jay)

F-14A-85-GR, **159618** (DR-17)
Left: This VF-31 F-14D(R) was photographed at NAS Miramar, Fightertown USA in October 1995. F-14Ds can be quickly identified by their turkey feathers, lack of glove vanes, Martin-Baker MB-14 ejection seats, and side-by-side mounting of the Television Camera System (TCS) and Infrared Search and Tracking Set (IRST). This Tomcat was delivered to VF-124 on 24 October 1975. It was assigned to VF-31 as of February 1998. (Author)

F-14A-85-GR, **159619** (DR-9)
Above: On 27 March 1994, VF-2 "stormed into Fallon," (squadron historian's words, not mine), for pre-cruise Air Wing strike training. This marked the first demonstration of the new carrier air wing concept comprising one Tomcat squadron and three F/A-18 Hornet squadrons. Prior to its remanufacture at Grumman, Bethpage, this Tomcat served with VF-124, VF-24, and VF-1. Following its conversion it was assigned to VX-4 where it underwent Electromagnetic Environmental Effects Testing. On 7 January 1994 it was transferred to the *Bounty Hunters* of VF-2. In December 1996 it was flying with VF-31. As of February 1998 this Tomcat was assigned to VF-213. (Grove)

F-14A-85-GR, **159620**
Left: Delivered to VF-124 on 12 November 1975, this Tomcat was transferred to VF-211 on 18 February 1976. For the rest of its service life this Tomcat alternated between VF-124 and VF-211. It was photographed in VF-211 markings 1 October 1987 on the transient ramp at NAS Willow Grove. It was stricken 12 October 1994 and is currently preserved at El Centro. (Roop)

F-14A-85-GR, **159621**
Right: Delivered to VF-124 on 24 October 1975, this VF-24 Tomcat was photographed in June 1981 wearing a gloss gull gray scheme. Originally known as the *Red Checkertails*, the squadron changed its name to the *Fighting Renegades* in the mid-1980s. This Tomcat was delivered to AMARC on 2 August 1994 and stricken 21 February 1995. (Author's Collection)

F-14A-85-GR, **159622**
Below: Armed with a single CATM-7 practice AIM-7 Sparrow, this VF-211 Tomcat was photographed at NAS Miramar 18 December 1976. The squadron's first F-14A flight occurred on 23 December 1975. Paired with VF-24, the unit made its first Tomcat cruise aboard the USS Constellation (CV-64) from 12 April to 21 November 1977. This Tomcat crashed on 15 July 1978 while flying from the USS Constellation. (Taylor via Sheets)

F-14A-85-GR, **159623**
Right: This VF-24 Tomcat was stricken 19 December 1981 while embarked aboard the USS Constellation (CV-64) in the Indian Ocean. Crewed by Cdr. Switzer and Ltjg. Baranek, its arrestor hook skipped the number three wire and snagged the four wire which had been incorrectly set for a lighter weight. The wire failed to bring the Tomcat to a halt, and both crewmen were forced to eject as their Tomcat went over the side. Both men were quickly rescued. (Spidle via Snyder)

F-14A-85-GR, **159624**
Left: The Navy's fade to gray continued through the 1980s as evidenced by this colorless VF-211 Tomcat photographed 14 August 1989. Delivered to VF-124 on 2 December 1975, this F-14A was transferred a week later to VF-211. It also flew with VF-2 before being retired to AMARC 5 December 1990. It was stricken 21 February 1995. (Anselmo)

F-14A-85-GR, **159625**
Above: This VF-24 Tomcat was photographed during May 1984. It was during this period the *Fighting Renegades* made their only cruise aboard USS Ranger (CV-61). This Tomcat was delivered to VF-124 on 11 November 1975. It served with VF-24, VF-2, and VF-213, and was delivered to AMARC on 31 January 1991. It was stricken 21 February 1995. (Trombecky/Airframe Images)

F-14A-85-GR, **159626**
Left: Delivered to VF-124 on 25 November 1975, this Tomcat was reassigned to VF-211 on 9 December 1975. It was photographed in their markings at NAS Fallon in June 1981. It later flew with VF-114, VF-213 and the Naval Weapons Fighter School prior to being stricken on 7 October 1994. This Tomcat is currently displayed at NAS Fallon in full-color VF-213 markings. (Grove).

F-14A-85-GR, **159627**
Right: Delivered on 1 December 1975, this Tomcat would see service with two squadrons, VF-124 and VF-24, by the end of the year. This Tomcat was also flown by VF-2 and VF-114 before returning to VF-124 in March 1992. It was stricken 22 September 1993. (Grove)

F-14A-85-GR, **159628** (DR-8)
Right: This VF-211 Tomcat was photographed 3 December 1977 following its first cruise aboard the USS Constellation (CV-64). This example also flew with VF-114 and VF-124 before entering NADEP Norfolk for remanufacture as an F-14D(R). Following service with VF-2 this Tomcat was flying with VF-31 as of June 1997. (Logan)

F-14A-85-GR, **159629** (DR-7)
Below: By the time this VF-24 Tomcat was photographed on 6 May 1978, the squadron had completed its first WestPac cruise aboard the USS Constellation. The silhouettes painted in red on this Tomcat's nose may depict interceptions of Russian aircraft sent to shadow the fleet. Initially delivered to the *Fighting Renegades* in December 1975, this Tomcat served with VF-124 prior to transfer to Grumman Bethpage for conversion to F-14D(R) standards. As of February 1998 it was assigned to VF-31. (Wilson)

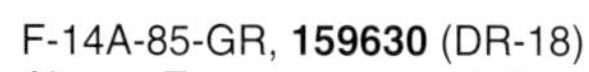
F-14A-85-GR, **159630** (DR-18)
Above: Even an overcast day can do little to detract from VF-2's high-vis paint scheme applied to this F-14D(R). It was photographed at NAS Oceana following the *Bounty Hunters* move there in April 1996. This Tomcat was delivered to VF-124 on 1 December 1975 and later served with VF-211 and VF-1 prior to entering NADEP for conversion to a F-14D(R) in February 1989. As of September 1997 it was assigned to VF-2. (Author)

F-14A-85-GR, **159631**
Left: This VF-24 Tomcat was delivered on 20 December 1975. The "9" surrounded by stars indicates this Tomcat was assigned to CVW-9 when it was photographed 20 November 1976. It also served with VF-114, VF-211, and VF-124 prior to being stricken on 8 October 1994. It currently rests at the San Diego Aerospace Museum Annex at Gillespie Field, CA. (Logan)

F-14A-85-GR, **159632**
Left: Delivered to VF-211 on 15 December 1975, this example was stricken on 25 August 1978 when it crashed into the Pacific Ocean 200 miles off the California coast. It served briefly with VF-1 during 1976-77. (Trombecky/Airframe Images)

F-14A-85-GR, **159633** (DR-16)
Right: After serving with VF-24, VF-211, and VF-213, this Tomcat was returned to Grumman emerging as the 16th F-14D(R). On 22 February 1994, while serving with VF-11 and flying from USS Carl Vinson (CVN-70), this Tomcat crashed approximately 900 miles southwest of San Diego, California. Both crewmen ejected and were rescued by a SH-60F from HS-8. (Dorr)

F-14A-85-GR, **159634**
Right: Photographed on 31 October 1976, this Tomcat was delivered to VF-211 on 22 December 1975. It served with VF-24, VF-154, and VF-213. Entering AMARC on 17 December 1990 it was stricken on 21 February 1995. (Logan)

F-14A-85-GR, **159635** (DR-15)
Below: When the *Tomcatters* of VF-31 moved to NAS Miramar 6 March 1992, they gave up their black-nosed F-14As for F-14Ds. This example, the 15th remanufactued F-14D(R), was rebuilt by Grumman Bethpage. It was stricken on 13 January 1995 when it was involved in a mid-air collision with F-14D 164340 over the Pacific Ocean south of San Diego. The crews were rescued by a SH-60F from HS-8, and a USMC CH-53. (Grove)

F-14A-85-GR, **159636**
Left: Delivered to VF-211 on 12 December 1975, this Tomcat was stricken on 26 November 1978 when it crashed in the Pacific Ocean 100 miles off the coast of Korea near Pusan. It served briefly with VF-1 during 1976-77. (JEM Slides)

F-14A-85-GR, **159637**
Below: This Tomcat was delivered to VF-124 on 17 December 1975. It served with VX-4 before being assigned to VF-211. It was captured on film in *Checkmates* markings at NAS Miramar in August 1988. It was later flown by VF-213, VF-114 and VF-124 before returning to VF-211 in 1993. This Tomcat was stricken on 19 September 1994. (Romano)

F-14A-90-GR, **159825**
Left: It is a welcome relief to see some color after looking at so many low-vis Tomcats. This Tomcat was delivered to VX-4 on 16 January 1976. It later served with VF-124, VF-2, VF-114, VF-213 and finished its career assigned to VF-124. It was photographed in VF-114 markings at NAS Miramar in May 1984. The 1000 modex was applied to commemorate the 1,000th arrested landing by Capt. Bob "Burner" Hickey, then CO of CVW-11. The event took place aboard the USS Enterprise (CVN-65) on 18 May 1984. This Tomcat was stricken 9 February 1994. (Trombecky/Airframe Images)

F-14A-90-GR, **159826**
Right: I seriously doubt I will ever track down an image of this Tomcat. It was modified by Grumman for spin recovery tests and delivered to NATC on 13 February 1976. It was stricken near Patuxent River on 5 March 1976 while conducting spin tests.

PHOTO NOT AVAILABLE

F-14A-90-GR, **159827**
Right: Photographed at Nellis AFB on 3 December 1976, this VX-4 *Evaluators* F-14A was wearing this splinter scheme as part of the Air Intercept Missile Evaluation (AIMVAL) and Air Combat Evaluation (ACEVAL) program held in 1976-77. Conducted over the expansive ranges of Nellis AFB, it was designed to test the combat effectiveness of the Navy F-14A and USAF F-15A. The evaluations confirmed time honored principles of aerial combat – speed is life! fuel is life! This example was delivered to VX-4 on 29 February 1976. It later served with VF-124, VF-1, VF-213, VF-102, and VF-101 where it remained through March 1996. (Logan)

F-14A-90-GR, **159828**
Below: Armed with a single AIM-54 Phoenix missile, this VX-4 Tomcat was photographed 6 September 1987. This example also took part in the AIMVAL/ACEVAL tests, which involved six brand-new Block 90 Tomcats. Later flown by VF-124, VF-301, VF-101, and NATC, it was stricken 19 January 1996. (Trombecky/Airframe Images)

F-14A-90-GR, **159829**
Left: Photographed at Patuxent River 20 August 1983, this former AIMVAL/ACEVAL Tomcat spent its entire operational career flying with test and evaluation units VX-4, PMTC, and NATC. As of February 1998 it was assigned to NAWC-AD at NAS Patuxent River. (Roop)

F-14A-90-GR, **159830**
Left: This VX-4 Tomcat was photographed at Nellis AFB 12 July 1977, during the AIMVAL/ACEVAL evaluations. As with 159827, this example displays a Keith Ferris splinter scheme. In the background are the F-15As which participated in the tests. Mounted on the chin of this Tomcat is a prototype version of Northrop's TCS (Television Camera System). This F-14A was stricken 31 March 1992 after completing its operational life with VF-124. It is currently on display at Jack Northrop Field, Hawthorne, CA. (Huston)

F-14A-90-GR, **159831**
Below: Delivered to VX-4 on 19 February 1976 and also used during AIMVAL/ACEVAL, this Tomcat was transferred to VF-202 in 1990 and again in 1993. During the interim it served with VF-124 and VF-301. It was stricken 27 September 1995 following service with VF-101. (Logan)

F-14A-90-GR, **159832**
Above: This rather drab VF-213 Tomcat was photographed at NAS Miramar during August 1988 following a WestPac/IO cruise aboard the USS Enterprise (CVN-65). Delivered to VF-124 on 15 March 1976, it was assigned to VF-213 on three different occasions. Other squadrons which flew this Tomcat were VF-24 and VF-154. In 1992 this Tomcat was at the Fleet Air Western Pacific Repair Activity Facility (FAWPRA) Cubi Point. It was stricken 22 February 1994. (Romano)

F-14A-90-GR, **159833**
Right: The *Aardvarks* of VF-114 began transitioning to the Tomcat on 15 December 1975. "Zot", the squadron's mascot, is a depiction of the aardvark from the comic strip B.C. This Tomcat, photographed 5 May 1984, contrasts sharply with 159834, a high-vis example. Following service with VF-21, VF-301, and VF-101 it was retired to AMARC 15 February 1996. (McGarry)

F-14A-90-GR, **159834**
Right: Photographed at NAS Fallon in June 1982, this is what an *Aardvarks* Tomcat should look like. For years, aircrew assigned to VF-114 scrounged for serviceable orange flight suits. Note the RIO properly attired in orange. Delivered to VF-124 on 15 March 1976, this example served with VF-2 and VF-114. It was delivered to AMARC on 21 March 1991 and stricken 9 August 1995. (Grove)

F-14A-90-GR, **159835**
Left: The Navy's fade to gray was nearly complete when this VF-24 Tomcat was photographed at NAF Washington, during November 1985. Delivered to VF-124 on 19 March 1976, this Tomcat was transferred to VF-1 by March 1979. It was stricken 16 September 1990. (Kopack)

F-14A-90-GR, **159836**
Left: This VF-1 Tomcat was fresh from the paint shop and in the process of having its squadron markings applied when photographed during August 1988. Delivered to VF-124 on 9 April 1976, it served with VF-213, VF-124, VF-114, and VF-101. As of September 1997, it was in SARDIP status at NAS Fallon. (Romano)

F-14A-90-GR, **159837**
Below: This *Aardvarks* Tomcat was photographed during VF-114's first cruise aboard the USS America (CV-66), 13 March to 22 September 1979. This well-traveled Tomcat also served with VF-124, VF-2, VF-51, VF-211, VF-154, VF-213, VF-1, and VF-24. It was delivered to AMARC on 7 July 1995. (Author's Collection)

F-14A-90-GR, **159838**
Above: Delivered to VF-124 on 2 April 1976, and coded NJ/426, it was stricken 21 June 1976 when it crashed near NAS Miramar following an engine explosion. (Trombecky/Airframe Images)

F-14A-90-GR, **159839**
Right: Delivered to VF-124 on 23 April 1976, this Tomcat crashed two days after the crash of 159838. Coded NJ/427, it too impacted near NAS Miramar. (Trombecky/Airframe Images)

F-14A-90-GR, **159840**
Below: This rather plain VF-124 Tomcat holds the distinction of being the 200th example delivered. That event occurred 26 April 1976. It was photographed at NAS Miramar 3 July 1976. It was stricken 18 January 1984, while serving with VF-114. This Tomcat reportedly lost power while attempting a single engine approach to the USS Enterprise. (via Sheets).

F-14A-90-GR, **159841**
Above: This VF-1 Tomcat, photographed on 10 February 1980, wears the "NE" tail code of CVW-2 and USS Ranger on its tail. Aside from a cruise aboard the USS Kitty Hawk (CV-63) during 1984, the *Wolfpack* would be embarked aboard the USS Ranger (CV-61) until the squadron's disestablishment 1 October 1993. This Tomcat last served with VF-41 and was stricken 29 November 1994. (Wilson)

F-14A-90-GR, **159842**
Left: This Tomcat crashed 19 April 1977 while taking part in ACM training near NAF El Centro. It reportedly suffered a double flame-out and departed controlled flight. It was photographed in VF-124 markings at Kirtland AFB, New Mexico, 27 February 1977. (LeBaron)

F-14A-90-GR, **159843**
Left: The *Aardvarks* of VF-114 emerged victorious from the ComFitAEWWingPac *High Noon* gun derby in 1982. The squadron also was named the 1981-82 "Mutha" award winners. Conceived by three VF-124 instructors in 1964, "Mutha" is awarded each year to the most operational and colorful squadron in the Pacific Fleet. This Tomcat, with appropriately attired RIO, was photographed during June 1982. Delivered to VF-124 on 19 May 1976, it served with VF-2, VF-114, and VF-213. It was stricken 20 July 1993 when it was involved in a ramp strike while coming aboard the USS Abraham Lincoln (CVN-72). Both crewmen, Lt. Matthew W. Claar and Lt. Robert D. Fuller ejected. However, Lt. Claar perished, and six crewmen aboard the ship were injured. (Grove)

F-14A-90-GR, **159844**
Right: The *Screaming Eagles* of VF-51 traded their F-4Ns for Tomcats in 1974. This example was photographed on 2 May 1992. It was delivered to the squadron in September 1989. Of interest are the aircrew callsigns – "Kong" and "Face." The first, given to Cdr. Bob King, is self evident. The reasoning for the other moniker is unknown. This Tomcat had previously served with VF-1 and VF-114 and later served with VF-213 before being stricken 3 October 1994. (Van Aken)

F-14A-90-GR, **159845**
Right: As a result of corrosion control efforts, this VF-213 Tomcat has taken on a mottled appearance common to the low-vis paint scheme. Initially delivered to NATC on 28 May 1976, this example was transferred to VF-124 on 30 December 1976. It later saw service with VF-2, VF-114, VF-213, VF-201, VF-24, and VF-32. As of February 1998 it was assigned to VF-101. (Roop)

F-14A-90-GR, **159846**
Below: Photographed sans any markings in November 1979, this Tomcat was delivered to VF-124 on 20 May 1976. It was transferred to VF-213 and was stricken on 14 February 1986 during the *Black Lions* WestPac-Indian Ocean cruise aboard the USS Enterprise (CVN-65). (Stewart)

F-14A-90-GR, **159847**
Above: Groundcrew perform post-flight checks on a VF-114 Tomcat at NAS Fallon. The squadron was taking part in predeployment carrier air wing training for its September 1982 to April 1983 cruise aboard the USS Enterprise (CVN-65). During the cruise the *Aardvarks* made over 250 intercepts of Soviet bombers and reconnaissance aircraft. This Tomcat saw service with VX-4, VF-124, VF-114, VF-211, VF-24, VF-111, and VF-2. Following delivery to NADEP North Island it was stricken 20 September 1990. (Grove)

F-14A-90-GR, **159848**
Left: By 15 October 1977 VF-101, the east coast Fleet Readiness Squadron, had assumed the role of training pilots for Atlantic Fleet Carrier Air Wings. At NAS Miramar, the *Gunfighters* of VF-124 were still busy training Pacific Fleet Tomcat crews and maintainers. This example, delivered to VF-124 on 16 August 1976, later saw service with VF-114, VF-1, VF-213, VF-302, VF-84, and VF-14. It is currently displayed at the Tilamook Naval Air Station Museum, Oregon. (Wilson)

F-14A-90-GR, **159849**
Left: Although the aircrews retained their orange flight suits, the color was gone from the majority of VF-114 Tomcats when this example was photographed 22 November 1986. Prior to entering AMARC 21 March 1995, this Tomcat had served with VF-124, VF-114, VF-111, VF-302, and VF-41. (Rys)

F-14A-90-GR, **159850**
Right: The *Red Lightnings* of VF-194 were re-established on 1 December 1986. They were scheduled to make their first operational cruise aboard the USS Independence (CV-62) in 1988. Budget cuts resulted in the disestablishment of the squadron 30 April 1988, prior to the cruise. This example was photographed one month before the squadron went away. In 1989 this Tomcat was reassigned to VF-51, then VF-1. Its career ended where it began, with VF-124. It was stricken 24 February 1995. (Snyder)

F-14A-90-GR, **159851**
Right: Delivered to VF-124 on 16 August 1976, this Tomcat was assigned to VF-213 early in 1977. It was stricken 25 March 1978 when it entered a spin and crashed into the Pacific Ocean near Cubi Point, the Republic of the Philippines. At the time it was coded NH/210, and was taking part in a WestPac cruise aboard the USS Kitty Hawk (CV-63). (Trombecky/Airframe Images)

F-14A-90-GR, **159852**
Below: Of interest are the "double nuts" atop the vertical stabilizer on this VF-2 CAG Bird. It would appear the *Bounty Hunters*, at least until 1980, staved-off the decree for low-vis paint schemes. Delivered on 18 August 1976, this example served with VF-124, VF-1, VF-114, VF-51, and VF-213 prior to being stricken 24 February 1995. (Author's Collection)

F-14A-90-GR, **159853**
Above: Will the Political Correctness Police be knocking on my door if I dare describe this Tomcat as VX-4's "Black Bunny?" This Tomcat is gone now, stricken on 10 September 1993. A black Tomcat still flies at Point Mugu, (164604), but it is referred to as "Vandy One." The Playboy Bunny has been replaced by VX-9's Vampire bat design. This Tomcat served with VF-124 and VF-213 prior to reaching VX-4 in 1979. On interest is the crying bunny applied temporarily for a May 1992 farewell photo shoot. (Roth via Kaston)

PHOTO NOT AVAILABLE

F-14A-90-GR, **159854**
Left: Delivered to VF-124 on 20 August 1976, this Tomcat suffered an engine fire and loss of control forcing the crew to eject 28 June 1977. When it was stricken this Tomcat was assigned to VF-114. Coded NH/101 it was flying a training sortie in restricted zone W-291, southwest of San Diego.

F-14A-90-GR, **159855**
Left: Long known for their interesting paint schemes, (just look at the F-16Ns in the background), NFWS gave their Tomcats the same treatment. In August 1991, *TOPGUN* acquired the first of four Tomcats. The type is used by instructors to fly wing in order to improve a students learning and to teach adversary pilots how to fly against the Tomcat. This was a Su-27 look-alike when photographed 14 December 1991. It was initially delivered to VF-124 on 22 September 1976. It later served with VF-213, VF-1, VF-111, VF-114, and VF-2. As of February 1998 it was assigned to the NSAWC NAS Fallon. (Grove)

F-14A-90-GR, **159856**
Above: This Tomcat was delivered to VF-124 on 3 September 1976. It was transferred to VF-114 in 1977 and assigned to VF-51 in 1989. On 16 March 1990, *Screaming Eagle* 106 and *Sundowner* 202, (160676), made the first Pacific Fleet intercept of a Soviet Tu-22M-3 Backfire. This example was photographed at NAF Washington, on 13 January 1990. This Tomcat also served with VF-301 and VF 101. It was stricken on 8 April 1996. (Author)

F-14A-90-GR, **159857**
Right: Photographed at NAF Washington, in April 1985, this VF-213 Tomcat was assigned to CVW-11. Although USS Enterprise (CVN-65) appears just aft of the glove vane, VF-213 and its brethren, VF-114, did not embark aboard this carrier during 1985. This example was assigned to VF-124 on 21 August 1976. It later served with VF-213, VF-1, VF-302, and VF-202. It was stricken 11 April 1994. (Author)

F-14A-90-GR, **159858**
Right: Photographed at NAS Miramar 12 February 1977, this VF-114 Tomcat is representative of paint schemes in the halcyon days of high visibility markings. The name USS Kitty Hawk, the *Aardvarks'* first assigned aircraft carrier, appears near the glove vane. Initially assigned to VX-4 on 31 August 1976, this example reached VF-114 later the same year. Before it entered AMARC on 23 August 1995, this Tomcat had served with VF-2, VF-124, VF-1, VF-24, and VF-101. (Wilson)

F-14A-90-GR, **159859**
Left: This VF-213 CAG bird was photographed 17 September 1978 at NAS Miramar. It was stricken 8 August 1984, during a sortie from NAS Cubi Point, Republic of the Philippines. This accident, as well as a number of others, prompted then Secretary of the Navy John Lehman to testify before the Congressional Appropriations subcommittee that the F-14 and its TF 30 turbofan were, “probably the worst engine/airframe mismatch we have had in many years. The TF30 is just a terrible engine and has accounted for 28.2% of all F-14 crashes.” (Logan)

F-14A-90-GR, **159860**
Above: The *Black Lions* fade to gray would seem complete when this example was photographed at McGuire AFB on 27 January 1987. Before reaching VF-213, this Tomcat had served with NATC, VF-114, and VF-124. It would also serve with VF-21, VF-302, and VF-101. It was stricken 4 March 1996. (Roop)

F-14A-90-GR, **159861**
Left: This Tomcat was stricken in these markings on 3 September 1986, almost ten years to the day after its delivery date. It was photographed 27 February 1982 at a USAF air show. (Stewart)

F-14A-90-GR, **159862**
Right: This Tomcat currently rests at AMARC. It was photographed at NAF Washington, in December 1986 while assigned to VF-114. Delivered on 30 September 1976 it was the first Tomcat assigned to VF-114. During the delivery flight from Grumman, this Tomcat suffered a massive failure of its hydraulic system. Cdr. Walter J. Davis, Jr. and Lt. Joseph Kumpan, Jr. made a successful landing at Tinker AFB. (Author)

F-14A-90-GR, **159863**
Right: Maybe colorful F-14s will make a comeback in the late 1990s. This colorful VF-14 Tomcat was photographed at NAS Oceana 20 October 1995. It holds the distinction of being the first Tomcat to fire Zuni rockets, a feat it performed on 31 August 1995. Tomcats are receiving Zuni capability due to its planned use in the Forward Air Control (FAC) mission. Before reaching the *Tophatters* this example was assigned to VF-301, VF-124, VF-2, VF-1, and VF-213. The latter squadron was the first to take delivery of it 15 October 1976. In December 1996 it was assigned to VF-14. By 18 April 1997 it was resting at AMARC. (Author)

F-14A-90-GR, **159864**
Below: Not another VF-114 Tomcat! Many of these Block 90 Tomcats only flew with one or two squadrons during their entire career. This example was delivered to VF-114 on 2 November 1976. It was transferred to VF-51 in 1992 and then to VF-213 in 1995. As of February 1998 this Tomcat was assigned to VF-201 and was in the process of being stripped of useful parts. (Richard)

F-14A-90-GR, **159865**
Left: This VF-124 Tomcat was photographed in 1988 at the zenith of low visibility markings. Initially delivered to VF-213 on 19 October 1976, this example served with VF-1, VF-2, and VF-124. This Tomcat collided with 160888 while rolling out after landing at NAS Alameda on 16 December 1988. It was stricken 3 May 1989. (Romano)

F-14A-90-GR, **159866**
Left: Delivered to the fleet on 27 October 1976, this VF-114 Tomcat was photographed 2 December 1981, the same year the squadron received the "Golden Tailhook" for recording the best carrier landing performances during a cruise aboard the USS America (CV-66). This example was delivered to AMARC on 1 May 1996. (Svendsen)

F-14A-90-GR, **159867**
Below: The *Black Lions* of VF-213 were paired with VF-114 from 1977 until the disestablishment of the *Aardvarks* in 1995. Photographed 6 October 1979, this example was stricken 22 February 1996 while assigned to VF-24. On a post maintenance test hop, this Tomcat crashed in the Persian Gulf not far from the USS Nimitz (CVN-68). The crewmen, LCdr. Roger Pyle and Lt. Thomas Eberhard ejected, suffering minor injuries. In light of this crash and that of 161158, on 18 February 1996, CNO Adm. Mike Boorda ordered a 72-hour safety stand-down for all F-14 Tomcats. (Wilson)

F-14A-90-GR, **159868**
Right: This *Aardvarks* Tomcat stands out even under an overcast sky. Delivered 12 November 1976, this example was later transferred to VF-124, VF-201, and VF-101. As of 1 October 1997 it was resting at AMARC. (Knowles)

F-14A-90-GR, **159869**
Below: This Tomcat was not being delivered to a NAS Dallas squadron when photographed there in 1980. It was on route to VF-213 following an overhaul. Of interest to scale model builders is the placement of the warning data and other miscellaneous stencils. Delivered to VF-213 on 30 October 1976, this F-14A later served with VF-2, VF-24, VF-111, and VF-124. It was stricken 9 September 1994. (Wilson)

F-14A-90-GR, **159870**
Right: Photographed at NAS Miramar 15 October 1977, this VF-114 Tomcat was about to go aboard the USS Kitty Hawk for the squadron's first blue water cruise. The USS Kitty Hawk (CV-63) was the first of four carriers to embark CVW-11 and VF-114. The USS America (CV-66), USS Enterprise (CVN-65), and USS Abraham Lincoln (CVN-72) were the others. This Tomcat was stricken on 8 July 1991. (Wilson)

F-14A-90-GR, **159871**
Above: The *Black Lions* of VF-213 received this Tomcat 21 November 1976. A year later the squadron was preparing for its first cruise, with CVW-11 and VF-114 aboard the USS Kitty Hawk (CV-63). Of interest are the twin tails on the squadron's lion mascot. This Tomcat was still flying twenty years later, assigned to NAWSC, NAS Fallon. On 16 September 1997 it was delivered to AMARC. (Paul)

F-14A-90-GR, **159872**
Left: This VF-114 Tomcat was stricken on 27 January 1989 when it crashed into a Cotton Field near Eloy, Arizona. That much is known for sure. Navy investigators report the crew lost consciousness after taking off their helmets and oxygen masks, and donning garrison caps for an inflight photo. It was reported in a Washington Post article dated 24 June 1995, that the crew was attempting to "moon" another aircraft when they passed out. Both crewmen perished in the crash. (Knowles)

F-14A-90-GR, **159873**
Left: Work-ups for a major cruise usually involve the entire air wing coming together at NAS Fallon. When photographed during May 1985, this VF-213 Tomcat was taking part in CVW-11 work-ups for a January 1988 WestPac/Indian Ocean cruise aboard the USS Enterprise (CVN 65). This Tomcat was delivered to VF-213 on 20 December 1976. It later flew with VF-24, VF-21, and VF-2. As of February 1998 it was once again assigned to NSAWC. (Grove)

F-14A-90-GR, **159874**
Above: This was the 50th and final block 90 Tomcat constructed. It was delivered to VF-114 on 2 December 1976. It only served with one other squadron, VF-124, prior to being stricken on 24 February 1995. It was photographed during June 1977 at NAS Miramar while assigned to VF-114. (Zerbe)

F-14A-05-GR, **160299**
Right: The first IIAF Tomcat made its inaugural flight on 5 December 1975 and was delivered on 24 January 1976. U.S. Navy crews delivered the aircraft from Calverton, Long Island, New York, to Khatami Airfield, near Isfahan, Iran. The flights were supported by USAF KC-135 tankers. The first thirty IIAF Tomcats, **160299/160328** were coded 3-6001 to 3-6030 in Iranian service. (Grumman)

F-14A-05-GR, **160300**
Below: The first two Iranian Tomcats were delivered to Khatami AB, Isfahan, Iran, on 24 January 1976. The closest Tomcat is **160299** and it was crewed by an Iranian pilot and Grumman WSO. **160300** is in the background, crewed by Capt. Mohamed Farharvar, IIAF with Hughes NFO, E.S. "Mule" Holmberg. (Grumman)

F-14A-05-GR, **160301**
Above: This lineup of eleven Iranian Tomcats was taken on the flight line at Kahtami AB, Isfahan, Iran. Also included in this shot is **160311, 160309, 160313** and **160304**. Of interest are the aerial refueling probes which have had their covers removed. (Grumman)

F-14A-05-GR, **160306**
Left: This Tomcat, destined for Iran, was photographed at Grumman's Calverton facility during April 1976. (Leslie via Rotramel)

F-14A-05-GR, **160318**
Left: This Iranian Tomcat was photographed in 1976 in full Iranian markings at Khatami AB, Isfahan, Iran. Of interest is the U.S. style ground power unit. (Bero via Hughes)

F-14A-05-GR, **160321**
Right: This Iranian Tomcat was photographed on the flight line at Khatami AB, Isfahan, Iran in 1976. Khatami was built specifically for training Iranian Tomcat crews. Another F-14 base, Shiraz, was located further south. Of interest are the hardened shelters which were capable of holding two Tomcats each. (Bero via Hughes)

F-14A-05-GR, **160325**
Right: Iranian Tomcats were refueled by USAF KC-135s during delivery flights to Iran. Iran also possessed its own aerial refueling capability. The long-range plan called for all Iranian Tomcats to be reconfigured to the USAF flying boom system of aerial refueling. This example, in full Iranian markings, is most likely being refueled by an Iranian tanker via the U.S. Navy probe and drogue system. (Grumman)

F-14A-05-GR, **160326**
Below: This IIAF Tomcat waits at Grumman's Calverton facility for its Navy flight crew. It was either flown directly to Khatami Airfield or arrived there via NAS Rota, Spain. Of interest are the U.S. delivery markings. (JEM Slides)

F-14A-05-GR, **160327**
Above: The 29th Iranian Tomcat was photographed on a delivery flight to Iran. This flight originated at Grumman's Calverton Facility then to Spain and finally over Turkey to Iran. Of interest is Mt. Ararat in the background. (Bero via Hughes)

F-14A-05-GR, **160328**
Left: This overall view of the shop floor at Grumman's Calverton facility shows H-29, **160327** in the foreground. Undergoing construction next to it is H-30, **160328**. In the background is H-28, **160326**. (Grumman via Drendel)

F-14A-15-GR, **160340**
Left: In the foreground is Grumman Fab number 251 and assigned USN bureau number 160395. Undergoing construction next to it is H-42, destined for delivery to the Imperial Iranian Air Force as 160340. (Grumman via Drendel)

F-14A-15-GR, **160344**
Right: Beautiful inflight study of Imperial Iran Air Force Tomcat. The side number 3-6046 was applied by the Iranians and is repeated on the tail. (via Robert Dorr)

F-14A-15-GR, **160361**
Below: These IIAF Tomcats were part of the second batch ordered by the IIAF. Of interest are the Iranian markings. The Tomcat in the foreground, 3-6063 is **160361**. In the background, 3-6052, is **160350**. The 7 in the circle on the vertical stabilizer indicates that both were assigned to the 7th Air Wing, Imperial Iranian Air Force. (Paul)

F-14A-15-GR, **160378**
Right and opposite top: The majority of IIAF Tomcats had been delivered by the end of 1977. The following year, political unrest began to take hold throughout Iran and by 1 April 1979, the Ayatollah Khomenei declared the country an Islamic Republic. Meanwhile, the eightieth and final IIAF Tomcat was at Grumman scheduled for conversion to the USAF boom and receiver type aerial refueling system. The conversion was never completed and this airframe was placed in storage at AMARC. In 1986 it was delivered to NADEP, North Island, brought up to Navy standards and issued to PMTC on 13 November 1987. It was photographed upon its delivery to PMTC, and later in NAWC markings. As of February 1998 it was assigned to the Weapons Test Squadron, Point Mugu. (Vasquez)

See previous entry.

F-14A-95-GR, **160379**
Left: This attractive VF-41 Tomcat was photographed 12 August 1978, following its first cruise aboard the new USS Nimitz (CVN-68) The *Black Aces* were one of the first squadrons (along with VF-84), to make the transition to the Tomcat under the guidance of VF-101, the Atlantic Fleet Replacement Squadron. This example, delivered 4 January 1977, served with VF-101, VF-143, and again with VF-101 prior to being transferred to the Navy's Fighter Weapons School in 1996. (Huston)

F-14A-95-GR, **160380**
Below: Paired with VF-41, the *Jolly Rogers* of VF-84 stood up with the Tomcat in mid-April 1977. The "AJ" tailcode, displayed by both squadrons, indicates assignment to CVW-8. VF-84 would remain assigned to this air wing until shortly before its disestablishment 1 October 1995. This example was delivered to VF-84 on 24 March 1977. It was stricken 3 May 1980 when it crashed while being launched from the Nimitz on "Gonzo Station" in the Arabian Sea. The Nimitz and CVW-8 were present as part of the response to the Iranian Hostage situation. (Roop)

F-14A-95-GR, **160381**
Right: A star and lightning bolt have adorned VF-33 aircraft since the Korean war. This example was photographed at McGuire AFB, 8 July 1992. This Tomcat served with VF-101, VF-41, VF-31, VF-33, and VF-84. It was stricken 26 August 1994 following a catastrophic engine fire. This Tomcat crashed in the Pamlico Sound, off the coast of North Carolina. The crew, consisting of Lt. Jeff Daus and Lt. Kevin Martin were recovered and treated for minor injuries at the MCAS Cherry Point N.C., hospital. (Roop)

F-14A-95-GR, **160382**
Right: The "Skull and Crossbones" markings have been displayed on Navy aircraft dating back to VF-17, a World War Two fighter squadron. The *Jolly Rogers* of VF-84 traced their lineage to 1 July 1955 and were known then as the *Vagabonds*. More about the Legend of the "Bones" can be found on page 212. This Tomcat, delivered 8 April 1977, served with VF-101 and again with VF-84 before serving with VF-74, VF-11, and VF-51. As of February 1998 it was assigned to NSAWC. (Paul)

F-14A-95-GR, **160383**
Below: Delivered to VF-41 on 1 February 1977, this example was stricken 3 November 1979. The Tomcat was reportedly on a low-level intercept of a USAFE F-111 when it crashed within sight of the USS Nimitz. The Tomcat's crew consisted of Cdr. D.J. Formo, squadron CO, and LtCdr. N. Delello, the squadron maintenance officer. (Picciani Aircraft Slides)

F-14A-95-GR, **160384**
Above: This VF-84 Tomcat was photographed at NAS Oceana during May 1982. Assigned to CVW-8. The *Jolly Rogers* embarked aboard two carriers, the USS Nimitz (CVN-68) and USS Theodore Roosevelt (CVN-71). During 1982, VF-84 and VF-41 conducted sea trials aboard the third Nimitz-class carrier, USS Carl Vinson (CVN-70). Hence, the markings aft of the intake. This Tomcat was delivered 2 May 1977. It was later assigned to VF-101, VF-84, VF-32, and was retired to AMARC on 31 August 1995. (Author)

F-14A-95-GR, **160385**

Left: Delivered to VF-41 on 17 May 1977, this Tomcat was stricken 27 May 1981. It was one of three Tomcats damaged or written off when struck by EA-6B, BuNo 159910. The Prowler, assigned to VMAQ-2, Det Y, was attempting to land when it crashed aboard the USS Nimitz on 26 May 1981. Among the thirteen fatalities were three deck personnel from VF-41. (Trombecky/Airframe Images)

F-14A-95-GR, **160386**
Left: The *Jolly Rogers* took delivery of this Tomcat on 24 May 1977. Transferred to VF-101, it returned to VF-84 in 1980. Transferred to VF-142 in 1987, it later served with VF-14 and again with VF-101. As of February 1998 it was assigned to VF-101. (Ostrowski)

F-14A-95-GR, **160387**
Above: Photographed during May 1978, this VF-41 Tomcat spent its entire service life with two squadrons, VF-41 and VF-101. On 13 June 1996 it was delivered to AMARC. (Paul)

F-14A-95-GR, **160388**
Right: The overall gloss gray scheme was used prior to the tactical paint scheme. It was the result of a directive issued 18 February 1977. Compare this scheme with that applied to 160387. This *Black Aces* Tomcat was photographed 22 April 1979, the day before the NAS Norfolk Open House. This Tomcat was stricken when it crashed while on approach to the USS Nimitz (CVN-68), 1 April 1980. (Author's Collection)

F-14A-95-GR, **160389**
Right: Delivered to VF-101 on 5 August 1977, this Tomcat did not reach VF-74 until January 1989. It was photographed 4 June 1989, arriving at the London Ontario International Air Show, North America's premier air show. Besides assignments to VF-101 and VF-74, this example saw service with VF-84, VF-41, VF-33, VF-211, VF-21, and VF-51. In 1995 it was transferred to VF-24. As of June 1997 it was assigned to VF-211. By September 1997 it was still at NAS Oceana but was nothing more than a hulk stripped of all useful parts. (Author)

F-14A-95-GR, **160390**
This page: On 19 August 1981, *Fast Eagle* 107, piloted by Lt. Larry "Music" Muczynski, and RIO, Lt. Dave Anderson, was flying off the coast of Libya about to take part in a live missile exercise when it was fired upon by one of a pair of Libyan Su-22 Fitters. Lt. Muczynski launched an AIM-9L Sidewinder, which downed one Su-22. The crew of *Fast Eagle* 102, 160403, was flying lead and downed the other Sukhoi with a AIM-9L. **160390** again made headlines on 25 October 1994. Lt. Kara Hultgreen, one of the Navy's first two female aviators assigned to an operational fighter squadron, VF-213, was at the controls on that date. Lt. Hultgreen and her RIO, Lt. Matthew P. Klemish were attempting to land aboard the USS Abraham Lincoln (CVN-72), when her Tomcat's port engine malfunctioned causing the F-14A to crash. Lt. Klemish ejected successfully, Lt. Hultgreen did not. Her body was recovered from 3,700 of feet water still strapped into her ejection seat. (Dorr and Jay) (PH1 Bob Shanks/USN)

F-14A-95-GR, **160391**
Above: The *Jolly Rogers* were disestablished 1 October 1995. Before VF-84 went away, the squadron starred in their second motion picture, *Executive Decision.* This example was used in the flying sequences and was photographed immediately after completion of filming in September 1995. At this point in its history the squadron was no longer assigned to an air wing and therefore it displays no tail code. This Tomcat was delivered to VF-84 on 18 July 1977. It served briefly with VF-31 before returning to the *Jolly Rogers.* As of February 1998 it was flying with VF-41. (Author)

F-14A-95-GR, **160392**
Right: Delivered to VF-41 on 13 August 1977, this example was stricken less than two months later on 3 October 1977. While being recovered aboard the USS Nimitz (CVN-68), it suffered a ramp strike and was written off.

F-14A-95-GR, **160393**
Below: This beautiful study of a VF-84 Tomcat was made on 14 October 1978. Delivered to VF-84 on 20 August 1977, it was transferred during November 1986 to VF-102. After serving with VF-14 it was delivered to AMARC 16 April 1996. (Ostrowski)

PHOTO NOT AVAILABLE

F-14A-95-GR, **160394**
Above and left: Photographed at NAS Fallon during October 1994, this VF-41 Tomcat wears special "Bombcat" markings and nose art denoting its role in developing the Bombcat's air-to-mud capabilities. This example was stricken, 22 May 1995 while operating from the USS Theodore Roosevelt (CVN-71). In addition to VF-41, it served with VF-84 and VF-142. This was the 250th Tomcat constructed for the Navy. (Grove)

F-14A-95-GR, **160395**
Left: Delivered to VF-41 on 27 September 1977, this Tomcat served with VF-142 and VF-32 prior to its transfer to VF-33. Photographed 19 May 1991 in *Starfighters* markings, this example later served with VF-102 and VF-84. Officially stricken 17 May 1995, it was delivered the following day to the Kalamazoo Air Museum, in Kalamazoo, Michigan. It is displayed there in full VF-84 markings. (Roop)

F-14A-95-GR, **160396**
Above: Following delivery to VF-84 on 3 June 1977, this Tomcat served with VF-101, VF-84, VF-142 and VF-14. It was photographed 9 December 1997 following its delivery to VF-201. The Navy needs *The Hunters* low time Tomcats, so the Squadron will convert to the F/A-18 Hornet during 1998. (Richards)

F-14A-95-GR, **160397**
Right: Where's the bureau number on this VF-32 Tomcat? The photographer must have gone right up to the engine access panel and written it down. This F-14A is a classic example of the low-vis paint schemes of the 1980s and 1990s. Delivered to VF-84 on 18 June 1977, this aircraft served with VF-101, VF-142, VF-32, and VF-102. As of June 1997 it was once again flying with VF-101. (Rys)

F-14A-95-GR, **160398**
Below: This VF-41 Tomcat was delivered on 30 June 1977. Later assigned to VF-101, it went to the Naval Air Rework Facility in 1985. It was stricken 16 September 1990. (Author's Collection)

F-14A-95-GR, **160399**
Left: *Starfighter* 203 was the first Tomcat to surpass 5,000 flight hours. This milestone was reached on 2 February 1993. The crew for this historic flight was Cdr. A. R. Reade and Lt. Chris Stubbs. This example was photographed at NAS Fallon in August 1991. Delivered to the Navy on 12 July 1977, it was flown to NAS Oceana by LCdr. Vincent Lesh and Ltjg. Gerald Donato, and turned over to VF-41. The *Black Aces* flew it for 2,683.5 hours and made 942 arrested landings. Following depot-level maintenance it served with VF-101, VF-33, and VF-102. After serving with VF-14 it was delivered to AMARC on 12 June 1996. (Grove)

F-14A-95-GR, **160400**
Below: Taxiing at NAS Fallon, this VF-41 Tomcat was taking part in CVW-8 work-ups for the *Black Aces* 1981 Med cruise aboard USS Nimitz (CVN-68). This example was later lost while serving with VF-101. On 30 August 1983 it collided with Tomcat 161430 during an ACM training sortie near the Virginia Capes. Following the crash new programs dealing with spatial disorientation and midair collision avoidance were incorporated into aircrew training. (Grove)

F-14A-95-GR, **160401**
Left: According to the Navy's new "brown water" policy, carriers will be dealing with an increasing number of coastal operations, such as *Operation Deny Flight,* over Bosnia. Deny Flight began in 1993 and provided air cover and close air support for UN Protection Forces. This VF-102 Tomcat was photographed aboard the USS America (CV-66) steaming in the Adriatic during October 1993. Delivered to VF-84 on 3 August 1977, this Tomcat was serving with VF-41 in January 1997. By September 1997 it had become the gate guard at the Fleet Area Control and Surveillance Facility, Virginia Capes, near NAS Oceana. (G. Kromhout)

F-14A-95-GR, **160402**
Right: This Tomcat was delivered to VF-101 on 26 August 1977, and later assigned to VF-41 in 1979. It was one of three Tomcats damaged or destroyed during a landing accident aboard the USS Nimitz on 26 May 1981. Later rebuilt, it was delivered to VF-101 in 1992. On 10 January 1996 it was delivered to AMARC. (JEM via Shields)

F-14A-95-GR, **160403**
Above and right: As the crew of *Fast Eagle* 107 (160390) was downing a Su-22 on 19 August 1981, the lead Tomcat, *Fast Eagle* 102, crewed by Cdr. Hank Kleeman and RIO Lt. Dave Venlet, were downing the second with a AIM-9L Sidewinder. Tragically, Capt. Kleeman was killed 3 December 1985 while landing his F/A-18A, BuNo 162435, at NAS Miramar. His Tomcat was photographed early in 1982 after the modex had been changed to 101 and the kill marking had been removed from the tail. The yellow one on the tail indicates the squadron won the Adm. Clifton award as the Navy's best fighter squadron for 1981. Also on the tail is the CNO's "S" for Aviation Safety and the Battle "E." As of February 1998 this Tomcat was assigned to VF-154. (Jay)

F-14A-95-GR, **160404**
Left: Photographed 20 October 1995 on what photographers call "a blueberry day", (clear with low humidity), this VF-14 prepares to launch from NAS Oceana on an ACM training sortie. Delivered to VF-41 on 14 September 1977, this Tomcat later served with NWEF, VF-101, VF-11, VF-124, and VF-14. This Tomcat was delivered to AMARC on 20 April 1996. (Author)

F-14A-95-GR, **160405**
Left: A low-vis *Jolly Rogers* Tomcat taxis out at NAS Fallon during April 1986 while CVW-8 was working up for a North Atlantic cruise aboard the USS Nimitz (CVN-68). Delivered to VF-84 on 23 September 1977, this Tomcat later flew with VF-11 before returning to VF-84 in 1985. In December 1996 it was assigned to VF-211. This Tomcat was delivered to AMARC on 25 July 1997. (Grove)

F-14A-95-GR, **160406**
Below: Thunderstorms are a fact of life at NAS Oceana. At first glance, this image appears to have been taken at that venue. This VF-32 Tomcat was actually photographed at NAS Fallon, where CVW-3 was preparing for a cruise aboard the USS Eisenhower (CVN-69). Of interest are the BRU-41 improved triple ejector racks (ITERs) attached to the Phoenix weapons rails indicative of the Tomcat's newly acquired air-to-ground mission. Delivered to VF-101 on 28 October 1977, this example served with VF-84, VF-41, and VF-33 before reaching VF-32 in 1993. This Tomcat was delivered to AMARC on 21 November 1997. (Grove)

F-14A-95-GR, **160407**
Above: Delivered on 26 October 1977, this Tomcat has only served with three squadrons, VF-101, VF-41, and VF-32. Displaying VF-101 markings, it was photographed on 14 October 1978 at NAS Oceana. As of September 1997 it was flying with VF-32. (Ostrowski)

F-14A-95-GR, **160408**
Right: When photographed on the transient ramp at McGuire AFB 10 October 1988, VF-41 had just completed a North Atlantic cruise. Delivered to VF-101 on 29 October 1977, this Tomcat also flew with VF-84. As of September 1997 it was assigned to the NSAWC, NAS Fallon, Nevada. (Roop)

F-14A-95-GR, **160409**
Right: Wearing a well worn and touched up gloss gray paint scheme, this VF-41 Tomcat was photographed during a January 1981 deployment to NAS Fallon. Delivered to VF-101 15 November 1977, this F-14A was transferred to VF-41 in December 1979. It would later serve with VF-101 again, and with VF-143 until stricken 12 September 1988. (Grove)

F-14A-95-GR, **160410**
Above: Delivered to VF-101 on 1 November 1977, this Tomcat was retired to AMARC 26 April 1995. During the intervening years, it served with VF-101, VF-41, VF-33, VF-11, VF-211, and VF-24. It was wearing low-vis VF-41 markings when photographed during April 1986. (Grove)

F-14A-95-GR, **160411**
Left: Color began to creep back into Tomcat paint schemes by the mid-1990s. This example from VF-14 was photographed at NAS Oceana during April 1994. It was assigned to that squadron as of September 1997. (Author)

PHOTO NOT AVAILABLE

F-14A-95-GR, **160412**
Left: Delivered to VF-101 on 16 December 1977, this Tomcat was stricken due to an engine fire over the Atlantic Ocean on 21 March 1978. At the time it was assigned to VF-101 and coded AD/111.

F-14A-95-GR, **160413**
Right: The creature on the tail of this Tomcat is a Tarsier, a vicious, monkey-like creature from the Philippines. The Tarsier was the mascot and nickname for VF-33 until they adopted the *Starfighters* moniker. The Tarsier appears on this VF-101 Tomcat as a tribute to VF-33. A number of VF-101 Tomcats display the markings of recently disestablished Tomcat squadrons. VF-33 went away on 1 October 1993. This Tomcat was assigned to AMARC on 29 May 1996. (JEM Slides)

F-14A-95-GR, **160414**
Above: A beautiful image of a *Jolly Rogers* Tomcat at NAS Fallon. The *Jolly Rogers* tail markings, VF-84 and this Tomcat, received worldwide recognition following the squadron's appearance in the 1980 motion picture *Final Countdown.* This example was delivered to VF-101 on 22 December 1977. From its delivery date through early 1993, this Tomcat served with VF-84 and VF-101. During March 1993 it was transferred to VF-33. It was stricken 23 March 1995 while assigned to VF-14. This Tomcat crashed 75 miles off the coast of Virginia. The crew of USAF Capt. Vance C. Bateman, and Lt. Jerry Seagle were rescued by a SAR helicopter from NAS Oceana. (Grove)

F-14A-100-GR, **160652**
Right: The *Freelancers* of VF-21 stood-up with the Tomcat following an official ceremony held at NAS Miramar 15 March 1984. This example was not received by the squadron until February 1991. It was photographed in July 1991, a month prior to a WestPac cruise aboard USS Independence (CV-62). Delivered to VF-124 on 30 December 1977, this Tomcat served with VF-111, VF-2 and VF-1 prior to its assignment with the *Freelancers.* It completed its flying days with VF-101 and was stricken 19 February 1994. (Snyder)

F-14A-100-GR, **160653**
Left: This Tomcat was delivered to VF-124 on 12 January 1978. By 15 June 1978 it had been stricken. It was reportedly lost off San Clemente Island while conducting Field Carrier Landing Practice (FCLP). LCdr. Stallings and Ltjg. Dalley perished in the crash. (Trombecky/Airframe Images)

F-14A-100-GR, **160654**
Left: The *Sundowners* of VF-111 began F-14A operations in 1978. This example was initially delivered to VF-124 18 January 1978. It was assigned to VF-111 on two separate occasions: December 1981 to August 1986 and January 1988 to December 1992. It was photographed 2 May 1991 while the *Sundowners* were paired with the *Screaming Eagles* of VF-51 aboard the USS Kitty Hawk (CV-63) This Tomcat later served with VF-2, VF-124, and VF-101. It was stricken 17 April 1995. As of May 1997 it was a stripped hulk parked on the north side of NAS Oceana. (Vasquez)

F-14A-100-GR, **160655**
Below: This VF-84 Tomcat was photographed during the making of *Executive Decision*, filmed in Autumn, 1995. The plot involved a hijacked Boeing 747 carrying enough nerve agent to wipe out the entire East Coast. The *Jolly Rogers* Tomcats, armed with AIM-54, AIM-9, and AIM-7 air-to-air missiles, were tasked with intercepting and shooting down the jumbo jet . For the movie, two Tomcats were marked as VF-84 CAG birds. The side number for 160655 is actually 207. When VF-84 was disestablished 1 October 1995, this Tomcat was transferred to VF-101 where it was assigned as of January 1997. On 27 May 1997 this Tomcat was delivered to AMARC and assigned storage code AN1K0104. (Skarbek)

F-14A-100-GR, **160656**
Right: This VF-124 Tomcat was photographed at NAS Miramar 28 April 1979. It was transferred to VF-111 during December 1981. Coded NL/200, it was written off 31 March 1985, after suffering a hydraulic failure returning to USS Carl Vinson (CVN-70). The crew ejected after an unsuccessful attempt to divert to a shore base. (McIntosh)

F-14A-100-GR, **160657**
Right: The *Screaming Eagles* of VF-51 took delivery of their first Tomcat on 16 June 1978. This example was delivered 29 December 1981. The squadron, paired with the *Sundowners* of VF-111, made the inaugural cruise aboard the then new USS Carl Vinson (CVN-70). This Tomcat later served with VF-124 and VF-24 and was stricken 14 September 1994. (McGarry)

F-14A-100-GR, **160658**
Below: Aircrew and crewchief confer on the Strike Aircraft Test Directorate (SATD) ramp following a test hop. This F-14A was designated an NF-14A on 1 October 1993. The new designation reflects its role as a test and evaluation platform. Delivered to VF-124, 9 March 1978, it has served at NATC/NAWCAD since 1982. In 1997 it was being utilized to test the Digital Fight Control System. This example was photographed 11 April 1988. (Author)

PHOTO NOT AVAILABLE

F-14A-100-GR, **160659**
Left: This Tomcat was delivered to VF-124 on 31 March 1978. While assigned to VF-124, it crashed at sea immediately after launching from the USS Ranger (CV-61) on 13 September 1978. Both crewmen were recovered with no injury.

F-14A-100-GR, **160660**
Left: The markings developed by VF-194 were simple yet eye-catching. The squadron went away when CVW-10 was disestablished due to budget cuts. This Tomcat was delivered to VF-124 on 13 April 1978. It was transferred to VF-111 in 1982 and reached VF-194 in January 1988. It later served with VF-51 and was stricken 16 July 1991. (Grove)

F-14A-100-GR, **160661**
Below: Initially delivered to VF-124 on 1 May 1978, this Tomcat was transferred to VF-51 during December 1982. It was photographed at NAS Miramar 5 May 1994. Of interest are the partially lowered Phoenix pallets. Reassigned to VF-124, this example was stricken 27 September 1994. (McGarry)

F-14A-100-GR, **160662**
Above: These markings were standard to VF-124 Tomcats during the 1970s. This example was delivered 23 April 1978. It was stricken 9 April 1983 when it crashed into the Pacific Ocean while conducting night cyclic ops aboard the USS Carl Vinson (CVN-70). Both crewmen perished. (Wilson)

F-14A-100-GR, **160663**
Right: When photographed in February 1986, the *Screaming Eagles* markings had evolved somewhat. Although still wearing a gloss gray paint scheme, the all black tails had disappeared. This Tomcat was delivered to VF-124 on 5 May 1978. Assigned to VF-51 from December 1981 to May 1986, it was later transferred to VF-211 in May 1986. It was involved in an accident aboard USS Kitty Hawk (CV-63) on 2 September 1986. Delivered to the Naval Air Rework Facility at North Island, it was stricken 17 February 1987. (Peterson via Jay)

F-14A-100-GR, **160664**
Right: This very pristine VF-111 Tomcat was photographed at Tinker AFB, 10 May 1982. Delivered to VF-124 on 22 May 1978, it was transferred to VF-111 on 31 December 1981. It later served with VF-21, VF-301, and the Naval Air Weapons Center-Aircraft Division. It was stricken on 2 March 1995. (Greby Collection)

F-14A-100-GR, **160665**
Above: This Tomcat was the first example delivered to VF-51. It arrived from Calverton, Long Island, New York on 16 June 1978, only a week before being displayed at this air show. Of interest is the overall gloss gray scheme in which this Tomcat was delivered. In the 1984 Paramount movie *Top Gun,* this Tomcat was the mount of Lt. Pete "Maverick" Mitchell and Lt. Nick "Goose" Bradshaw. This example later served with VF-124 and VF-154. It was transferred to VF-101 and was still assigned to the *Grim Reapers* as of September 1997. (McGarry)

F-14A-100-GR, **160666**
Left: The inscription which accompanied this image reads "one millionth flight hour milestone aircraft, 26 March 1987." This VF-111 Tomcat was delivered with 160665 on 16 June 1978. It was transferred to VF-21 in 1989 and later to PMTC in December 1991. It has remained at Point Mugu through February 1998. (Huston)

F-14A-100-GR, **160667**
Left: Delivered to the *Screaming Eagles* on 6 July 1978, this Tomcat later served with VF-1 in 1985, and returned to VF-51 in 1986. It was later transferred to VF-213. As of June 1997 it was serving with the *Black Lions.* By February 1998 it was reassigned to NSAWC, NAS Fallon. (Huston)

F-14A-100-GR, **160668**
Right: By 1988 the tactical paint scheme had reached the *Bounty Hunters* of VF-2. This example is undergoing maintenance and is minus its port engine. It was delivered to VF-111 on 13 July 1978. Serving briefly with VF-2 during 1988, it was back with the *Sundowners* by February 1990. It finished its service life with VF-213 and has rested at AMARC since 16 March 1995. (Anselmo)

F-14A-100-GR, **160669**
Right: The paint schemes on VF-111 Tomcats went through many transitions. This example, photographed at NAF Washington 25 March 1989, is representative of one of the least colorful. Prior to reaching the *Sundowners,* this Tomcat flew with VF-51, VF-124, and VF-24. As of February 1998, it was assigned to NSAWC, NAS Fallon. (Author)

F-14A-100-GR, **160670**
Below: This *Sundowners* Tomcat was photographed 20 June 1980. It was delivered to VF-111 on 1 August 1978. On 3 November 1980, while being towed across the hangar deck of USS Kitty Hawk (CV-63), this Tomcat, with tractor in tow, went over the side and was lost off the coast of California. (McGarry)

F-14A-100-GR, **160671**
Above: This beautiful *Screaming Eagles* "CAG bird" was photographed 9 September 1979 at NAS Miramar. Of interest is the lower case 100, which follows the bureau number. This indicates that this is a block 100 F-14A. This Tomcat was delivered 1 August 1978. It served with VF-124 before transfer to the Naval Air Rework Facility in March 1986. It was next assigned to VF-191, until that squadron's disestablishment in April 1988. It later served with VF-124, VF-1, and VF-21. As of September 1997, it was assigned to VX-9 Det. Point Mugu. (Huston)

F-14A-100-GR, **160672**
Left: Delivered to VF-111 on 21 October 1978, this Tomcat was stricken less than a year later on 8 September 1979, when it crashed near Cubi Point, the Republic of the Philippines. At the time it was part of CVW-15 taking part in a WestPac cruise aboard the USS Kitty Hawk (CV-63). When written off, this Tomcat was coded NL/203 and assigned to VF-111. (Trombecky/Airframe Images)

F-14A-100-GR, **160673**
Left: This example was delivered to VF-51, 23 September 1978 and flew with VF-1, VF-124, and VF-111. It was later assigned to the Navy Fighter Weapons School/NSAWC, NAS Fallon. On 17 September 1997 it was delivered to AMARC and assigned storage code ANIK0108. It is shown here on 27 September 1997. (Patrick)

F-14A-100-GR, **160674**
Above: This *Sundowners* Tomcat taxies with wings swept as it is directed to a catapult aboard the USS Kitty Hawk (CV-63) 28 April 1979. This example was stricken on 27 June 1981 while operating in the Arabian Sea. This Tomcat was in a landing pattern when it developed smoke in the cockpit, erratic control imputes and warning lights. Command ejection was initiated by the RIO. The aircrew were recovered with minor injuries. (Huston)

F-14A-100-GR, **160675**
Right: This VF-51 Tomcat was photographed on 14 March 1981. While assigned to VF-124, and coded NJ/430, it was stricken 12 September 1988. (Wilson)

F-14A-100-GR, **160676**
Right: The *Sundowners* of VF-111 made their first Tomcat blue water cruise aboard the USS Kitty Hawk (CV-63) from 30 May 1979 to 25 February 1980. The cruise was extended when Iran took a number of U.S. citizens hostage. This example was photographed 22 March 1979. It went to NADEP, North Island in September 1994 and as of March 1996 was still listed at that location. (Huston)

F-14A-100-GR, **160677**
Left: This VF-51 Tomcat was delivered to the *Screaming Eagles* on 23 September 1978. It was photographed at Williams AFB during November 1980. While taking part in a WestPac cruise, it was knocked overboard the USS Kitty Hawk (CV-63) when struck by a landing A-7E assigned to VA-22. (Jay)

F-14A-100-GR, **160678**
Below: Delivered to VF-111 on 21 October 1978, this Tomcat served with VF-124 and again with VF-111 before reaching the *Black Knights* of VF-154. This Tomcat was later transferred to VF-213. As of February 1998 it was assigned to NSAWC, NAS Fallon. (Puzzullo)

F-14A-100-GR, **160679**
Left: This eye-catching Tomcat of VF-51 was photographed at NAF Washington during January 1979. Delivered 2 November 1978, it was transferred to VF-101 on 27 January 1995. It was delivered to AMARC on 4 December 1996. (Author's Collection)

F-14A-100-GR, **160680**
Above: Delivered to VF-111 on 28 October 1978, this Tomcat served twice with the *Sundowners*, from October 1978 to August 1986, and from December 1994 until 31 March 1995 when VF-111 was disestablished. This Tomcat was stricken on 16 August 1995 (Grove)

F-14A-100-GR, **160681**
Right: The 300th Navy Tomcat was delivered to VF-51 on 7 November 1978. During the filming of the movie *Top Gun*, this F-14A carried the names Lt. Tom "Iceman" Kazansky and Ltjg. Ron "Slider" Kerner on its canopy rails. It would later serve with VF-1 and VF-301. It was photographed in a NAS Miramar hangar on 18 August 1985. As of February 1998 it was assigned to VF-101. (Kaston)

F-14A-100-GR, **160682**
Right: The *Screaming Eagles* were assigned to the USS Carl Vinson (CVN-70) from 1983 through July 1990. The squadron was the first to utilize the Automatic Carrier Landing System and the first squadron to intercept Tu-22M-3 *Backfire* bombers. This Tomcat, photographed aboard the Vinson in January 1989, was delivered to VF-111 on 19 November 1978. It served briefly with VF-124 before assignment to VF-51 during April 1983. Later assignments included VF-154, VF-114, VF-101, and VF-124. As of 31 May 1996, this Tomcat was resting at AMARC. (Author's Collection)

F-14A-100-GR, **160683**
Left: This VF-51 Tomcat was photographed on 23 May 1979 during the squadron's first Tomcat cruise aboard the USS Kitty Hawk (CV-63). In 1983 this example was delivered to North Island's Naval Air Rework Facility. It was not returned to squadron service and was stricken on 22 October 1986. (Markgraf via Vasquez)

F-14A-100-GR, **160684**
Left: Retired to AMARC on 26 September 1994, this Tomcat currently is displayed with the ever expanding collection of the Pima County Air Museum. Delivered to VF-111 on 11 December 1978, and photographed 22 March 1979, this Tomcat later served with VF-124 and VF-51. It last served with VF-124 prior to delivery to Pima via AMARC. (Huston)

F-14A-100-GR, **160685**
Below: Delivered to VF-51 on 8 December 1978, this Tomcat was photographed 28 April 1979 while working up for a 1979-80 WestPac/Indian Ocean cruise. It appeared in the movie *Top Gun* with 160681 and 160695. The *Screaming Eagles* supported the movie with four Tomcats, eight aircrew and thirty-two maintenance personnel. While filming the aerial sequences, sixty-five sorties were flown in eleven days from NAS Miramar and NAS Fallon. This Tomcat was stricken 26 March 1986 while operating from the USS Carl Vinson (CVN-70). (Huston)

F-14A-100-GR, **160686**
Right: The *Sundowners* took delivery of this Tomcat on 2 December 1978. In 1989 it was reassigned to VF-111's sister squadron, VF-51. In May 1994 it was transferred to VF-211. As of September 1997 it was based at NAS Fallon, assigned to the NSAWC. (Huston)

F-14A-100-GR, **160687**
Below: Usually these camouflage schemes were water based and removable. This VF-14 Tomcat appears to have a permanent scheme applied. As of September 1997 this Tomcat was assigned to VF-32. Following service with VF-32 this Tomcat was delivered to AMARC on 9 October 1997. (JEM Slides)

F-14A-100-GR, **160688**
Right: Photographed at NAS Miramar 22 March 1979, this *Sundowners* Tomcat was delivered to VF-111 on 6 January 1979. It served with VF-124, VF-154, and VF-213 before being stricken 13 September 1994. (Huston)

F-14A-100-GR, **160689**
Left: This very toned down VF-124 Tomcat was delivered on 14 January 1979. It served with the *Gunfighters* until 1984 when it was transferred to VF-51. It later flew with VF-191, VF-111, and VF-24. As of September 1997 it was flying with the Fighter Weapons School/NSAWC at NAS Fallon. (Trombecky/Airframe Images)

F-14A-100-GR, **160690**
Left: Photographed at NAS Dallas 17 April 1982, this Tomcat was delivered to VF-111 on 6 November 1980. It had previously served with VF-124, commencing 19 January 1979. It reverted back to VF-124's control in 1989. It was stricken 19 September 1996 while assigned to VF-101. (Wilson)

F-14A-100-GR, **160691**
Below: This VF-124 Tomcat was part of the flight demonstration at the 1982 Dayton Air Show. Delivered 7 February 1979, it was transferred from VF-124 to VF-51 in 1986. Following a period with VF-21, during 1991-94, it reverted back to VF-51. Later assigned to VF-101, it was in storage at Grumman St. Augustine when stricken on 6 March 1996. (Henderson)

F-14A-100-GR, **160692**
Right: Photographed on NAS Willow Grove's transient ramp 25 November 1989, this VF-51 Tomcat has received the low-vis treatment. Gone are the red stripes on black stabilizers. Delivered from VF-124 in 1986, it was later transferred to VF-21 in 1991 and then returned to VF-51. As of September 1997 It was assigned to VF-101, the Atlantic Fleet Replacement Squadron. (Roop)

F-14A-100-GR, **160693**
Right: By late 1979, Grumman was delivering Tomcats in this interim overall gloss gray scheme. This example, assigned to VF-124, was photographed 8 September 1979. It was transferred to VF-24 in 1984. It would later serve with VF-2, VF-1, and VF-302. As of June 1997, it was flying with VF-101. (Wilson)

F-14A-100-GR, **160694**
Below: Photographed at Reese AFB during August 1985, this VF-51 Tomcat wears the remnants of its *Top Gun* movie markings. The fictitious VF-1 badge is actually the insignia for VAW-110, a NAS Miramar E-2C Hawkeye unit. During the movie *Top Gun*, this Tomcat displayed Lt. Mitchell's name and callsign plus that of his new RIO, Ltjg. Sam "Merlin" Wells. It was also equipped as a camera ship and shot many of the dramatic air-to-air ACM footage. This Tomcat was stricken 26 September 1994 and is currently on display in Corpus Christi, TX. (McMasters)

F-14A-100-GR, **160695**
Above: Photographed 15 June 1991, this Tomcat was assigned to VF-2. It was later transferred to VF-213. This Tomcat has also flown with VF-124, VF-24, VF-1, and VF-51. It was stricken on 29 February 1996. As of September 1997 it was in SARDIP status at NAS Fallon. (Kaston)

F-14A-100-GR, **160696**
Two photos at left: This was the 315th Tomcat and the final block 100 delivered to the Navy. It was assigned to VX-4, on 16 June 1979, and utilized testing the Tactical Air Reconnaissance Pod System (see TARPS image below left). Of interest are the extended glove vanes. This Tomcat later served with VF-111, VF-124, VF-2, VF-154, and VF-211. As of May 1998 it was assigned to VF-101. (Grumman)

F-14A-105-GR, **160887**
Right: This rather vanilla VF-24 Tomcat was photographed at NAS Fallon in May 1981. This example was delivered to VF-124 in April 1979 and was involved in a rather unusual accident on 28 November 1979 when it experienced an unexpected ejection by the RIO, Ltjg. Matheney. The pilot landed safely at NAS North Island but his RIO was lost at sea. Following service with VF-24 it was transferred to VF-1. It was stricken 22 January 1992 following an inflight fire. The crew failed to eject. (Grove)

F-14A-105-GR, **160888**
Below: The *Grim Reapers* of VF-101 took receipt of this Tomcat on 30 April 1979. It would later serve with VF-24 before being assigned to VF-124 in 1985. This Tomcat collided with 159865 while rolling out after landing at NAS Alameda on 16 December 1988. It was stricken 9 May 1990. (Van Dam)

F-14A-105-GR, **160889**
Right: Delivered to VF-124 on 1 May 1979, this Tomcat served with VF-24 from 1980 to 1989. It is pictured at NAS Fallon during June 1981 in VF-24 markings. It was returned to VF-124 and finished its flying career with VX-9 at Point Mugu. It was stricken 12 July 1995 and transferred to the Museum of Flying at Santa Rosa, in California's Napa Valley. (Grove)

PHOTO NOT AVAILABLE

F-14A-105-GR, **160890**
Left: Delivered to VF-124 on 19 May 1979, this Tomcat was stricken on 12 October 1979 when it crashed into the Pacific Ocean following departure from controlled flight during an ACM engagement. The aircrew ejected and were rescued.

F-14A-105-GR, **160891**
Below: This VF-74 Tomcat is just starting to retract its landing gear after departing a runway at NAS Oceana on 29 April 1988. It was delivered to VF-101 on 11 May 1979 and transferred to VF-74 in June 1983. Aside from a period at NADEP, it remained with the *Bedevilers* until April 1990 when it was transferred to VF-101. As of February 1998 it was serving with NSAWC, NAS Fallon. (Author)

F-14A-105-GR, **160892**
Left: This VF-51 Tomcat was photographed from the front to emphasize the Television Camera System, gun, pitot tube, inlet and AOA sensors. The TCS beneath the nose aids in acquiring and identifying aerial targets. This Tomcat now rests at AMARC, having been delivered there in 26 April 1995. Assigned to VF-124 on 6 June 1979, it also served with VF-114, VF-24, VF-51, and VF-211. (Author)

F-14A-105-GR, **160893**
Above: Photographed in April 1994, this *Black Aces* Tomcat wears a very mottled interim paint scheme, no longer high-vis and not as toned down as the tactical scheme. It was delivered to VF-101 on 12 June 1979. It later served with VF-143. From 1983 to 1992 it was assigned to VF-101 and then VF-103. Following service with VF-41 this Tomcat was delivered to AMARC on 14 November 1997. (Greby)

F-14A-105-GR, **160894**
Right: The *Black Knights* of VF-154 transitioned to the Tomcat in 1984, receiving twelve F-14As including three modified for TARPS. This example was photographed at NAF Washington in March 1991. The squadron was paired with VF-21 and currently forward deployed to Japan. It was delivered to VF-124 on 21 June 1979 and later served with VF-114 and VF-24 before assignment to VF-154. This Tomcat was transferred to VX-9 in 1995 and stricken 27 February 1996. (Author)

F-14A-105-GR, **160895**
Right: Delivered to VF-124 on 26 June 1979, this Tomcat was stricken on 9 September 1981. While assigned to VF-114 it crashed into the Indian Ocean following loss of control. The crew were rescued without injury. (Crimmins)

F-14A-105-GR, **160896**
Above: The *Swordsmen* of VF-32 spent October, November, and December 1994 taking part in *Operation Southern Watch* over Iraq. This VF-32 Tomcat was photographed aboard the USS Eisenhower (CVN-69) in January 1995 while participating in *Operation Deny Flight*. This Tomcat was delivered to VF-101 on 5 July 1979. It later served with VF-32 and VF-74, being reassigned to VF-32 in May 1989. As of February 1998 it was flying with VF-101. (Maglione)

F-14A-105-GR, **160897**
Left: Records indicate this Tomcat has only served with two squadrons, VF-14, the squadron it was assigned to when it was photographed on 24 January 1994, and VF-74. The *Tophatters* delivered this Tomcat to AMARC on 24 April 1996. (Grove)

F-14A-105-GR, **160898**
Left: The crew of this *Black Aces* Tomcat gave the photographer a thumbs-up as they taxied out for a training sortie armed with practice CTAM-9 Sidewinders. When this example was photographed in April 1990, VF-41 and their sister squadron, VF-84, had just returned from a two month deployment to test the deckcrew and systems aboard the USS Abraham Lincoln (CVN-72). This Tomcat was delivered to VF-32 on 1 September 1979. It later served with VF-142, VF-101, and VF-103. It was stricken 9 December 1995. (Rogers)

F-14A-105-GR, **160899**
Right: The arrow on this VF-103 *Sluggers* Tomcat is a carryover of the markings applied to the squadron's F-4S Phantoms. When this example was photographed in June 1983, the unit had just completed the transition to the F-14A. Originally delivered to VF-14 on 7 September 1979, this Tomcat would later fly with VF-103, VF-101, VF-33, and VF-102. It was stricken 22 February 1995. (Author)

F-14A-105-GR, **160900**
Above: This VF-32 Tomcat, equipped with a TACTS pod and a practice CTAM-9 Sidewinder, was photographed landing at NAS Oceana in September 1995. It was delivered to the squadron on 7 September 1979. It served with VF-142 and VF-74 prior to its return to the *Swordsmen* in 1992. As of 1 October 1997 it was resting at AMARC. (Author)

F-14A-105-GR, **160901**
Right: This VF-32 Gulf War veteran currently rests at AMARC, delivered there on 20 March 1996. It was photographed at NAS Oceana 27 March 1991, returning from *Operation Desert Storm*. Of interest is the appearance of blue paint, a result of corrosion control efforts. This example was delivered to VF-14 on 7 September 1979. It served with VF-74 from 1983 to 1989 prior to joining VF-32. (Author)

F-14A-105-GR, **160902**
Above: This VF-41 Tomcat, photographed during June 1996, is representative of the latest in a long line of *Black Aces* markings. Compare this scheme with VF-41 Tomcats 160379, 160381, 160408, and 160893 in order, to fully comprehend this evolution of this squadron's markings. This example was delivered to VF-32 on 28 September 1979. It saw service with VF-101, VF-142, VF-14, VF-84, VF-41, and VF-101. In January 1998 it was delivered to Calverton for display there. (Author)

F-14A-105-GR, **160903**
Left: Delivered to VF-14 on 29 September 1979, this Tomcat also served with VF-103, VF-101, and VF-41. It was photographed during October 1981 following its transfer from CVW-1 to CVW-6. It was stricken 15 October 1995. (Grove via Logan)

F-14A-105-GR, **160904**
Left: This *Sluggers* Tomcat was photographed at NAS Fallon during air wing work-ups for CVW-17's history-making 1985 MED cruise aboard the USS Saratoga (CV-60). During that cruise the *Sluggers* took part in the capture of the *Achille Lauro* hijackers. This image is a very good example of the demarcation between the different grays of the tactical paint scheme. Delivered to VF-14 on 21 February 1980, it later served with VF-103, VF-101, VF-33, and VF-84. It arrived at AMARC 31 August 1995. (Grove)

F-14A-105-GR, **160905**
Right: Photographed at NAS Oceana following its delivery on 28 September 1979, this Tomcat has yet to receive full VF-32 *Swordsmen* markings. It later served with VF-142, VF-74, VF-11, and VF-24. It was stricken on 27 February 1996. (McIntosh)

F-14A-105-GR, **160906**
Right: "Clifton 80" painted on the side of this VF-32 Tomcat indicates it was the Navy's best fighter squadron in 1980. The CNO's Safety "S" was awarded for completion of ten years and 33,000 hours of accident free flying. Delivered to the *Swordsmen* on 18 October 1979, this Tomcat later served with VF-74, VF-14, VF-102, and VF-101. It was stricken 1 March 1996. (Grove)

F-14A-105-GR, **160907**
Below: The *Tophatters* of VF-14 took delivery of this Tomcat on 18 October 1979. It was photographed during October 1981, wearing the toned-down interim paint scheme. It was stricken 20 September 1982, during a MED cruise aboard the USS Independence (CV-62). Reportedly, it crashed into the Mediterranean Sea 120 miles from Naples. (Author's Collection)

F-14A-105-GR, **160908**
Left: Delivered to VF-32 on 20 November 1979, this Tomcat was photographed while assigned to VF-74, at NAS Oceana in October 1984. Of interest is the bluish cast to this low-vis paint scheme. During 1989, this example was assigned to VF-84. Following service with VF-14 this Tomcat was delivered to AMARC on 18 December 1997 and assigned storage code ANIK0116. (Author)

F-14A-105-GR, **160909**
Left: This VF-14 Tomcat was photographed at McGuire AFB 15 July 1989. It carries a practice Sidewinder and is equipped with TCS. This example served with VF-14, VF-74, and VF-101. As of February 1998 it was being flown by VF-101. (Roop)

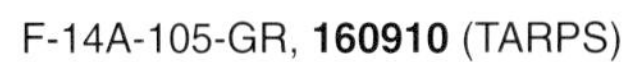

F-14A-105-GR, **160910** (TARPS)
Below: This VF-213 Tomcat was photographed during March 1985. Of interest is the canopy which appears to have come from another Tomcat. This Tomcat was one of fifty modified on the production line to carry TARPS. Three older F-14As were so modified by Norfolk's rework facility. Delivered to VF-124 on 19 November 1979, this example served with VF-213, VF-211, and VF-111. As of February 1998 it was assigned to VF-101. (Author's Collection)

F-14A-105-GR, **160911** (TARPS)
Above: Following delivery to VF-124 on 21 November 1979, this Tomcat served with VF-213, VF-302, and VF-201. While assigned to VF-302, it took part in the 1990 Reconnaissance Air Meet at Bergstrom AFB. A time exposure was required to capture this image on film. A flash was prohibited because it would have ruined the aircrew's night vision. As of June 1997 this Tomcat was flying with VF-101. On 27 August 1997 it was delivered to AMARC and assigned storage code ANIK0106. (Author)

F-14A-105-GR, **160912**
Right: This VF-14 Tomcat was photographed in October 1981. The slash under the Battle "E" indicates the squadron has received this award for combat readiness two years in a row. Delivered to the *Tophatters* on 29 January 1980, this example later served with VF-103, VF-101, and VF-11. It was stricken on 8 September 1994. (Author's Collection)

F-14A-105-GR, **160913**
Right: The *Tomcatters* of VF-31 the second oldest Navy squadron, possesses the most highly recognized mascot in the form of "Felix the Cat." With its trademark black nose,and red tails, VF-31 Tomcats were often referred to as flying pencils. This example was photographed 28 July 1990. Delivered to VF-32 on 25 June 1980, this Tomcat also served with VF-103, before its assignment to VF-31 in 1989. When the *Tomcatters* transitioned to the F-14D in 1992, it was transferred to VF-211. As of February 1998 it was assigned to NSAWC at NAS Fallon. (Anselmo)

F-14A-105-GR, **160914** (TARPS)
Above: When photographed 11 March 1980, this VF-124 Tomcat was participating in Photo Derby '80. This Tomcat was delivered to VF-124 on 6 December 1979. In September 1996 it was flying with VF-213. As of February 1998 it was assigned to VF-101. (Linn)

F-14A-105-GR, **160915** (TARPS)
Left: This TARPS-equipped Tomcat was delivered to VF-124 on 15 December 1979. It was transferred to VF-211 during December 1980. It served with VF-154, VF-124, VF-213, VF-111, and was transferred to VF-124 a third time in December 1992. As of September 1997 it was assigned to VF-154. (Puzzullo)

F-14A-105-GR, **160916** (TARPS)
Left: This Tomcat was photographed on 9 January 1980, during a stop at El Paso International Airport while on its delivery flight to VF-124. It was written off in a crash at NAS Miramar on 3 March 1980. During its takeoff roll, this Tomcat impacted an arresting gear engine. Both crew ejected and were recovered without injury. (McMaster)

F-14A-105-GR, **160917**
Above: This Tomcat may end its flying career with VF-14, the squadron it began flying with on 10 January 1980. Between assignments with the *Tophatters* it flew with VF-74 and VF-101. It was photographed at NAS Oceana on 20 October 1995 and was assigned to VF-14 as of February 1998. (Author)

F-14A-105-GR, **160918**
Right: Preceded by its shadow, this VF-41 Tomcat skims across the Saudi Arabian desert during *Operation Desert Storm*. This example was delivered to VF-32 on 15 January 1980. It later served with VF-74, and VF-101, before reaching the *Black Aces* of VF-41. It was stricken 26 October 1995. (USN)

F-14A-105-GR, **160919**
Right: Hauling a "doomsday" loadout of AIM-54 Phoenix missiles, this much published image of a VF-32 Tomcat has been seen by just about everyone. Delivered to VF-32 on 26 January 1980, it later served with VF-103, VF-31, VF-24, and VF-124. As of December 1996 it was in storage at NAS Jacksonville. (Grumman)

F-14A-105-GR, **160920** (TARPS)
Left: This VF-213 Tomcat was the mount of Lt. Davey Jones when it was photographed during June 1980 while taking part in air wing work-ups for a cruise aboard the USS Enterprise (CVN-65). Delivered to VF-213 on 30 January 1980, it served with VF-211, VF-213, VF-154, and VF-111. As of February 1998 it was flying again with VF-211. (Grove)

F-14A-105-GR, **160921** (TARPS)
Above: Delivered to VF-213 on 7 February 1980, this TARPS-equipped Tomcat was photographed at NAS Fallon taking part in Photo Derby '82. It would later serve with VF-154 until entering the Naval Aviation Depot, North Island in 1986. It was stricken 16 September 1990. (Linn)

F-14A-105-GR, **160922**
Left: This VF-11 Tomcat was photographed 22 July 1990. It was delivered to VF-14 on 4 February 1980 and served with VF-103, VF-101, VF-11, VF-24, and VF-124. It was stricken 24 February 1990. (Anselmo)

F-14A-105-GR, **160923**
Right: By 1986, the *Jolly Rogers* appeared to have made the fade-to-gray. This low-vis Tomcat has a hard demarcation between the two shades of gray similar to 160908. It was delivered to VF-32 on 11 February 1980, and served with VF-101, VF-84, VF-31, and VF-24. It was stricken on 6 March 1996. (Author)

F-14A-105-GR, **160924**
Right: Delivered to VF-14 on 14 February 1980, this Tomcat later served with VF-101, VF-41, and VF-33. It was photographed in VF-41 markings at NAS Fallon preparing for an October 1986 cruise to the North Atlantic. This Tomcat was stricken 4 September 1992. (Grove)

F-14A-105-GR, **160925** (TARPS)
Below: This TARPS Tomcat was delivered to VF-213 on 5 March 1980. It later served with VF-154 and VF-124. When the *Gunfighters* were disestablished, it was transferred to VF-101. Following a second tour of duty with VF-213 it was delivered to VF-201 where it remained through January 1998. (Grove)

F-14A-105-GR, **160926** (TARPS)
Left: On 28 December 1990, VF-84 and the rest of CVW-8 headed to the Persian Gulf aboard USS Theodore Roosevelt (CVN-71). This TARPS-equipped Tomcat was flying a Tactical Aerial Reconnaissance sortie when photographed late in *Operation Desert Storm*. This example was delivered to VF-124 on 11 March 1980. It served with VF-101 and VF-84. As of September 1997 it was serving with VF-41. (Morgan)

F-14A-105-GR, **160927**
Above: The "68" on the island of this aircraft carrier indicates this VF-41 Tomcat was photographed aboard the USS Nimitz (CVN-68). The *Black Aces* and CVW-8 were taking part in a North Atlantic cruise when photographed in September 1986. This Tomcat was stricken 10 September 1995 following service with VF-32, VF-101, VF-41, VF-102, and VF-101. It was delivered to AMARC on 8 September 1995. (Author's Collection)

F-14A-105-GR, **160928**
Left: This VX-4 "Bombcat" was photographed during July 1994. Of interest are a pair of dumb bombs hanging on the forward Phoenix pallets. Although normally associated with missile tests, VX-4 and later VX-9 have also participated in tests exploring the Tomcat's air-to-mud capability. This F-14A was delivered to VF-14 on 27 March 1980. It served with VF-101, VF-74, VF-41, VF-11, and VF-1. As of February 1998 it was assigned to VX-9 Det. Point Mugu. (Grove)

F-14A-105-GR, **160929**
Above: This VF-32 Tomcat currently rests at AMARC. Delivered to VF-32 on 24 March 1980, it later served with VF-101, VF-41, VF-31, and VF-211. It was photographed at NAS Fallon during October 1981. (Grove)

F-14A-105-GR, **160930** (TARPS)
Right: This low-vis VF-2 Tomcat was photographed 2 May 1992 at NAS Miramar. It was delivered to VF-124 on 8 April 1980. It later served with VF-211, VF-154, VF-2, and VF-124. As of March 1996 it was assigned to VF-213. (Van Aken)

F-14A-110-GR, **161133**
Right: Although this Tomcat displays VF-1 tail markings, it is assigned to VF-101. The Grim Reapers have painted a number of Tomcats as a tribute to recently disestablished Tomcat squadrons. This Tomcat was delivered to VF-101 on 9 May 1980. As of February 1998 it was in SARDIP status at Point Mugu. (Author)

F-14A-110-GR, **161134** (TARPS)
Left: "Moon Equipped" painted on the nose of this VF-101 Tomcat is not an indication it has been modified as a hot-rod, nor is it an advertisement for racing equipment. The name on the canopy rail, "Cdr. Moon Vance," is the reason for the nose art and moon on the vertical stabilizer. This example was photographed at NAS Oceana on 1 May 1982. It was a regular performer at many east coast air shows. Delivered to VF-101 on 1 May 1980, it served with VF-103, VF-84 and, as of September 1997, was flying with VF-41. (Roop)

F-14A-110-GR, **161135** (TARPS)
Above: Photographed at NAS Oceana, October 1981, this VF-101 Tomcat sports a variation of the splinter scheme applied to a number of VX-4 Tomcats used during AIMVAL/ ACEVAL. This TARPS modified Tomcat also displays a false canopy applied to cause visual deception during ACM. Delivered to VF-101 on 1 May 1980, it later served with VF-31, VF-102, and VF-32. As of September 1997 it was flying with VF-32. (Author)

F-14A-110-GR, **161136**
Left: This rainy shot was taken at Andrews AFB prior to the 1982 air show. Delivered to VF-101 on 23 May 1980, it later served with VF-41, VF-11, and VF-124. It was stricken on 24 February 1995. (Author)

F-14A-110-GR, **161137** (TARPS)
Above: This TARPS-modified Tomcat was photographed on 23 March 1996. It is interesting because it depicts the markings of VF-84, including its last CO's name, six months after the squadron was disestablished. It served with VF-84 from 1990 until the squadron's disestablishment 1 October 1995. These markings were retained by VF-101 to commemorate VF-84's history. In 1994, Capt. Snodgrass reported as the CO of Fighter Wing Atlantic Fleet, hence his name was retained on the canopy rail. As of 22 May 1996, this Tomcat was resting at AMARC. (JEM Slides)

F-14A-110-GR, **161138** (TARPS)
Right: This fully-marked VF-84 Tomcat was photographed in February 1981 following the *Jolly Rogers* starring role in the motion picture *Final Countdown*. This Tomcat was written off aboard the USS Nimitz (CVN-68) 27 May 1981 in the same EA-6B landing accident involving VF-41 Tomcats 160385 and 160402. Two VF-84 Plane Captains died as a result of this flight deck accident. (Jay)

F-14A-110-GR, **161139**
Right: This VF-32 Tomcat was photographed 11 January 1995 during CVW-3's first cruise aboard the USS Eisenhower (CVN-69). This Tomcat was delivered to VF-101 on 31 June 1980 and remained with the *Grim Reapers* until transferred to VF-32 in 1995. As of June 1997 it was still assigned to the *Swordsmen*. (Bottaro)

F-14A-110-GR, **161140** (TARPS)
Left: This VF-102 Tomcat was photographed during *Operation Deny Flight*, aboard the USS America (CV-66). Loaded with two AIM-9 Sidewinders, two AIM-7 Sparrows, a pair of external fuel tanks, and a TARPS pod, it is being prepared for an armed reconnaissance sortie over Bosnia. Delivered to VF-84 on 17 June 1980, this example served with VF-101, VF-32, and VF-102. As of 2 October 1997, this Tomcat was in storage at AMARC. (G. Kromhout)

F-14A-110-GR, **161141** (TARPS)
Left: There is only one location where this VF-84 Tomcat could have been photographed: Cairo, Egypt. At the time, the *Jolly Rogers* were taking part in a *Bright Star* exercise with the Egyptian Air Force and USAF. This example was delivered to VF-84 on 27 June 1980. It later served with VF-31 prior to being transferred to VF-211 in 1992. As of February 1998, it was flying with VF-154. (Jay)

F-14A-110-GR, **161142**
Below: During the winter of 1981, VF-33 held a paint scheme contest. The winning submission came from AQ2 Steve Mudgett. The only problem was that the *Tarsiers* had not yet received a Tomcat to apply the scheme to. Cdr. David Frost, the Commanding Officer of VF-101, permitted VF-33 personnel to apply the scheme to this *Grim Reapers* Tomcat, hence the "AD" tail code. This Tomcat was stricken on 3 March 1995. As of May 1997 it was a stripped out shell parked on the north side of NAS Oceana. (Author)

F-14A-110-GR, **161143** (TARPS)
Right: Delivered to VF-41 on 7 July 1980, this Tomcat was stricken 29 July 1982 while serving with VF-101. It crashed shortly after take-off from NAS Oceana. It was photographed at NAS Oceana on 1 May 1982. (Roop)

F-14A-110-GR, **161144** (TARPS)
Above: This Tomcat was delivered to VF-124 on 17 October 1980. On 26 September 1988 while assigned to VF-111 this Tomcat experienced and inflight fire while operating in the Arabian Sea. The crew was forced to eject. (Grove)

F-14A-110-GR, **161145**
Right: Delivered to VF-101 on 11 August 1980, this Tomcat later served with VF-14 and again with VF-101. It was stricken 8 September 1994. (Kaminski)

F-14A-110-GR, **161146** (TARPS)
Left: This VF-211 CAG bird is identified as a reconnaissance capable Tomcat by (TARPS) appearing after its bureau number. The squadron was recently awarded the CNO's "S" for safety and "Battle E." Delivered to VF-124 on 7 November 1980, this Tomcat later served with VF-154, VF-211, and VF-111. It was stricken on 20 September 1995 while flying with VF-213. The Tomcat's pilot, Lt. Neal P. Jennings and RIO, Ltjg. Timothy J. Gusewelle, were on a training mission approximately 55 miles from the USS Abraham Lincoln (CVN-72) when they were forced to abandon their craft. Both crewmen were rescued by the USS John Paul Jones (DDG-53). The Lincoln was steaming about 800 miles from Guam at the time of the crash. (Svendson)

F-14A-110-GR, **161147** (TARPS)
Above: This VF-102 Tomcat was photographed aboard the USS America (CV-66) during *Operation Deny Flight.* Initially delivered to North Island's Naval Air Rework Facility, it was assigned to VF-31 during January 1981. It served with VF-102 from 1985 until its transfer to VF-41 in 1994. As of June 1997 it was still flying with the *Black Aces.* (G. Kromhout)

F-14A-110-GR, **161148**
Left: When photographed at NAS Oceana in October 1985, this recently repainted Tomcat had just been transferred from VF-101 to VF-33. It was stricken while flying with VF-33, from the USS America (CV-66) on 23 August 1986. (Author)

F-14A-110-GR, **161149** (TARPS)
Right: This Tomcat was delivered to the Naval Air Rework Facility, North Island on 17 September 1980. Transferred to VF-31 on 18 February 1981, it crashed off the coast of Lebanon on 11 November 1983. It was photographed at NAS Oceana in full VF-31 markings, in October 1981. (Author)

F-14A-110-GR, **161150** (TARPS)
Right: On 25 September 1980, this Tomcat was delivered to the Naval Air Rework Facility, North Island. The *Tomcatters* of VF-31 took delivery on 25 February 1981. Arguably, this paint scheme is not suited for aerial combat. Regardless, VF-31 managed to retain colorful markings longer than most squadrons. This example later served with VF-102, VF-101, VF-32, and VF-84. It was stricken 14 September 1994. (Author)

F-14A-110-GR, **161151**
Below: The "AB" tail code indicates this VF-33 Tomcat was assigned to CVW-1 when photographed during October 1986. Although it displays a star on its vertical stabilizer, the squadron did not become officially known as the *Starfighters* until 1987. Prior to that, they were referred to as the *Tarsiers* and/or *Starfighters.* The squadron took part in *Operation Prairie Fire* and *Operation El Dorado Canyon* in March and April, 1986. Delivered to VF-101 on 8 October 1980, this Tomcat also served with VF-31 prior to being stricken 8 April 1994. (Richard)

F-14A-110-GR, **161152** (TARPS)
Left: Delivered to VF-124 on 2 October 1980, this Tomcat was photographed at NAS Oceana on 11 October 1981. It later served with VF-111, VF-124, and VF-302. As of December 1996 this TARPS-equipped Tomcat was flying with VF-201. (Roop)

F-14A-110-GR, **161153** (TARPS)
Left: Delivered to VF-124 on 2 October 1980, this *Sundowners* Tomcat was taking part in a WestPac/Indian Ocean cruise aboard the USS Carl Vinson (CVN-70) when it was photographed during December 1986. It was stricken on 20 September 1987 when it departed the flight deck when the #4 arrestor cable failed. (Van Aken)

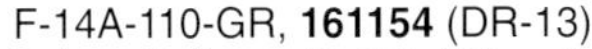

F-14A-110-GR, **161154** (DR-13)
Below: Delivered to VF-101 on 7 November 1980, this Tomcat was returned to Grumman during November 1991 for conversion to a F-14D(R). Assigned to VF-11 in March 1993, it was stricken a month later on 13 April 1993. (Author's Collection)

F-14A-110-GR, **161155** (TARPS)
Above: This Tomcat was delivered to the Naval Air Rework Facility at North Island on 7 November 1980. It was assigned to VF-143 in March 1981. It served with VF-101, VF-103, and VF-102. It was photographed in VF-102 markings aboard USS America (CV-66) while making an Italian port-of-call during CVW-1's 1991-1992 Med/Red Sea/Persian Gulf cruise. Following service with VF-32 this Tomcat was delivered to AMARC on 1 October 1997. (Maglione)

F-14A-110-GR, **161156** (TARPS)
Right: This VF-143 Tomcat was taking part in pre-cruise work-ups at NAS Fallon when photographed in April 1981. Of interest is the blackened area surrounding the 20mm M61-A1 cannon. Following the bureau number appears (TARPS). This denotes it as one of the sixty-six F-14As wired to carry the reconnaissance pod. This example was delivered to the Naval Air Rework Facility, North Island on 3 December 1980. It later served with VF-143, VF-101, VF-103, VF-84, and VF-32. As of 16 November 1996 it was resting at AMARC. (Grove)

PHOTO NOT AVAILABLE

F-14A-110-GR, **161157**
Right: Delivered to VF-101 on 2 December 1980, this Tomcat was stricken 24 April 1981 when it crashed on take-off from NAS Oceana.

F-14A-110-GR, **161158** (DR-3)
Left: Delivered to VF-32 on 1 May 1980, this example later served with VF-84 and VF-124 prior to returning to Grumman for remanufacture to F-14D(R) standards in 1990. It emerged as the third F-14D(R) and was assigned to VF-11 in 1992. On 18 February 1996, while flying approximately 120 miles west of San Diego, the Tomcat's starboard engine caught fire. The crew, Lt. Terence Clark and Cdr. Lewis "Scooter" Lamoreaux perished in the crash. (Author)

F-14A-110-GR, **161159** (DR-1)
Left: The *Swordsmen* of VF-32 took delivery of this Tomcat on 25 April 1981. Photographed at NAS Fallon during April 1983, it displays the *Gypsy* Tomcat markings designed by VF-32's Corrosion Control. This F-14A was returned to Grumman's Bethpage, New York facility in May 1990. It emerged as the first F-14A remanufactured to F-14D(R) standards. Following service with VF-31 it was transferred to VF-213 where it remained through May 1998. (Grove)

F-14A-110-GR, **161160**
Below: As of February 1998, this Tomcat was flying with the *Black Aces* of VF-41. When it was photographed at NAS Oceana, 17 April 1995, it was assigned to VF-101 and displayed VF-114 commemorative tail markings. It was delivered to VF-101 on 23 January 1981. (JEM Slides)

F-14A-110-GR, **161161** (TARPS)
Right: This VF-102 Tomcat was photographed preparing to launch from Bergstrom AFB, 22 August 1988. The *Diamondbacks*, along with the *Grim Reapers* of VF-101, were representing the Atlantic Fleet at the 1988 Reconnaissance Air Meet. Sixteen teams participated in the competition. Each team had to fly seven low-level, high speed missions. This TARPS-equipped Tomcat was delivered to NATC 9 February 1981. It later served with VF-101 and VF-102. In January 1997, this Tomcat was flying with VF-41. On 20 May 1997 it was delivered to AMARC and assigned storage code AN1K00103. (Author)

F-14A-110-GR, **161162** (TARPS)
Above: This VF-32 Tomcat was photographed at NAS Oceana on 23 September 1996. It was delivered to VF-32 on 16 May 1981 following modification to TARPS configuration by the Naval Air Rework Facility at North Island. Since September 1995, VF-32 has been instrumental introducing Digital TARPS to the fleet. As of February 1998 it was still serving with VF-101. (Kaston)

F-14A-110-GR, **161163** (DR-11)
Right: This F-14D(R) of VF-2, was photographed at NAS Willow Grove in October 1993. Delivered to VF-143 on 2 March 1981, it later served with VF-11, VF-101, and NATC before returning to Grumman. It emerged from the remanufacture process as the eleventh F-14D(R). Following remanufacture it was reassigned to VF-31. As of February 1998 it was flying with VF-101. (Sagnor)

F-14A-110-GR, **161164** (TARPS)
Left: This VF-84 Tomcat was photographed aboard the USS Roosevelt (CVN-72) during Operation Provide Comfort. A TARPS Tomcat, it was delivered to VF-143 on 3 March 1981. It later served with VF-101, VF-103, VF-84, and VF-32. As of February 1998 it was assigned to VF-211. (PHC Dennis Keske/USN)

F-14A-110-GR, **161165** (TARPS)
Below: This VF-124 Tomcat was photographed at NAS Fallon while participating in Photo Derby 1982. A predecessor to the Reconnaissance Air Meet, Photo Derby pitted RF-4B/Cs, RF-8Gs, and TARPS-equipped F-14As against each other in a series of tactical reconnaissance missions involving air-to-air and ground-to-air threats. This Tomcat was delivered to VF-124 on 6 March 1981. Transferred to VF-211 in April 1981, it was stricken on 4 September 1984 when the crew ejected near NAS Miramar following an in-flight fire. (Linn)

F-14A-110-GR, **161166** (DR-6)
Left: Delivered to VF-143 on 19 March 1981, this *Pukin' Dogs* CAG-bird was photographed at NAS Fallon during May 1983. Capt. Tom Treanor, Commander of the Air Group, appears on the canopy rail. This Tomcat later flew with VF-11, VF-101, VF-142, and with VF-11 again before delivery to the NADEP Norfolk during March 1991. It emerged in January 1994 as the Navy's sixth F-14D(R). Initially delivered to VF-2 it was later assigned to VF-11. As of February 1998 it was once again flying with VF-2. (Grove)

F-14A-110-GR, **161167** (TARPS)
Right: The bureau number of this VF-211 Tomcat reads 161167 110 (TARPS). This means it is a block 110 Tomcat with TARPS capability. Delivered to VF-124 on 25 March 1981, this Tomcat was photographed at NAF Washington on 23 June 1984. It was stricken 13 August 1986. (Zorn)

F-14A-110-GR, **161168** (TARPS)
Right: This VF-211 Tomcat carries the names of the *Flying Checkmates*' CO and XO on its canopy rail. It was photographed launching from NAS Miramar on 26 February 1989. Delivered to VF-124 on 8 April 1981, it was assigned to VF-211 twice, the first time in 1986 and then again in 1994. By February 1998, it was assigned to VF-101. (Trombecky/ Airframe Images)

F-14A-115-GR, **161270** (TARPS)
Below: This TARPS-equipped Tomcat was delivered to VF-124 on 23 April 1981. It served briefly with VF-111 prior to returning to VF-124 in 1985. Following a visit to the Naval Air Rework Facility at North Island, it was issued to VF-194. It was photographed at NAS Miramar wearing *Red Lightning* markings on 12 March 1988, just prior to the squadron's disestablishment on 30 April 1988. This Tomcat was later issued to VF-124 and then VF-154. As of June 1997 it was flying with the *Black Lions* of VF-213. (Snyder)

F-14A-115-GR, **161271** (TARPS)
Above: This VF-2 Tomcat was photographed aboard the USS Ranger (CV-61) returning from the Gulf War in June 1991. The Catapult Officer or "shooter" is a split-second away from launching this Tomcat. As of February 1998 this F-14A was assigned to VF-211. (Anselmo)

F-14A-115-GR, **161272** (TARPS)
Left: The *Evaluators* of VX-4 were the first squadron to operate the Tomcat. This example was photographed at Point Mugu 10 April 1988. By 1990, VX-4 had become heavily involved evaluating the F-14D Tomcat software, Naval Aircrew Common Ejection Seat (NACES), the F-14's air-to-ground capabilities, and by August, support of *Operation Desert Shield/Desert Storm.* Initially delivered to VF-124 on 12 May 1981, this Tomcat was transferred to VF-111 prior to entering North Island's rework facility in 1986. Post rework, it was assigned to VX-4, VF-111, and VF-213. Following service with the *Black Lions* it was assigned to VF-154. It remains with the *Black Knights* as of February 1998. (Trombecky/Airframe Images)

F-14A-115-GR, **161273** (TARPS)
Left: This *Black Lions* Tomcat was stricken 27 April 1995 when it went out of control and crashed. An investigation revealed the F-14A was banking at "an inappropriate angle" when an engine stalled and the Tomcat entered a irrecoverable flat spin. The crew was able to abandon the stricken craft prior to impact. The pilot, LCdr. Stacey Bates, was not so fortunate when he was flying 162599 on 29 January 1996. (Author's Collection)

F-14A-115-GR, **161274**
Right: This VF-1 CAG-bird wears the interim, toned down paint scheme when it was photographed at NAS Fallon during June 1983. This particular Tomcat holds the distinction of being the 400th example delivered to the Navy. The *Wolfpack* accepted it on 28 May 1981. It later saw service with VF-154, VF-213, and VF-24. As of February 1998 it was flying with VF-211. (Grove)

F-14A-115-GR, **161275** (TARPS)
Above: This VF-2 Tomcat was delivered 11 June 1981. It was photographed at Bergstrom AFB later the same month. Transferred to the North Island rework facility in 1990 it was later issued to VF-154. As of February 1998 it was in storage at NAS Jacksonville. (Author's Collection)

F-14A-115-GR, **161276** (TARPS)
Right: The *Bounty Hunters* of VF-2 took delivery of this Tomcat on 25 June 1981. In 1993 it was transferred to VF-213. As of February 1998 it was assigned to VF-211. This example was photographed at NAF Washington during October 1983. (Taylor)

F-14A-115-GR, **161277** (TARPS)
Left: This Tomcat was delivered to VF-101 on 2 July 1981. It served with VF-143 and again with VF-101. Transferred to NADEP Jacksonville in 1992, it has remained there through December 1996. (Crimmins)

PHOTO NOT AVAILABLE

F-14A-115-GR, **161278** (TARPS)
Left: It is extremely doubtful an image of this Tomcat will ever surface (no pun intended). While on a pre-delivery test flight, this Tomcat crashed into the Atlantic Ocean 12 miles off the coast of Long Island, New York. Power in one engine was lost while the Tomcat was at a high angle of attack. Then the second engine lost power. Neither engine could be restarted and the crew ejected.

F-14A-115-GR, **161279**
Below: As of February 1998, this Tomcat was assigned to the *Vampires* of VX-9 Det Point Mugu. VX-9 was formed at NAWS China Lake, following the disestablishment of VX-4 and VX-5 in 1994. The Point Mugu Det retained the "XF" tailcode mixed with the *Vampires* tail markings. This Tomcat was delivered to VF-1 on 2 August 1981. It later served with VF-191, VF-21, and VF-24, joining VX-9 late in 1994. (Author)

F-14A-115-GR, **161280** (TARPS)
Right: This VF-102 Tomcat was photographed returning to NAS Oceana 17 April 1991, following *Operation Desert Storm*. During the Gulf War, the *Diamondbacks* logged more than 1,400 combat hours flying a variety of missions. This Tomcat was delivered to VF-101 on 22 August 1981. It has served with VF-31, VF-103, VF-102, and VF-101. As of February 1998 it was in storage at NAS Jacksonville. (Author)

F-14A-115-GR, **161281**(TARPS)
Below: A pair of VF-101 Tomcats banks away from the camera ship somewhere over Virginia on 20 October 1982. The closest example is a TARPS version which was delivered to the *Grim Reapers* on 21 August 1981. It later served with VF-143, VF-103, VF-102, VF-101, and VF-32. As of February 1998 this Tomcat was serving with NSAWC, NAS Fallon. (Lawson)

F-14A-115-GR, **161282**(TARPS)
Right: This VF-101 Tomcat was photographed during the 1988 McGuire AFB open house and air show. During its service life this F-14A was assigned to VF-143, VF-32, VF-31, and VF-211. It holds the distinction of being the last F-14A flown by VF-31 prior to the *Tomcatters* departure for NAS Miramar and transition to the F-14D. This Tomcat was transferred to VF-154 and stricken 18 May 1996. It crashed into the Pacific Ocean 500 miles west of Guam after suffering an engine failure. The pilot Lt. Nigel S. Aitkins and RIO Lt. Edward F. Schmitt ejected safely. (Author)

F-14A-115-GR, **161283**(TARPS)
Above: This *Diamondbacks* Tomcat appeared at the NAS Norfolk open house in 1982. It was stricken on 20 June 1984 when it slid off an elevator aboard USS America (CV-66). (Author)

F-14A-115-GR, **161284**
Left: This VF-124 Tomcat was photographed at NAF Washington on 28 May 1988. It has served with VF-1, VF-124, VF-51, VF-21, VF-154, and VF-213. As of February 1998, this Tomcat was in storage. (Author)

F-14A-115-GR, **161285**(TARPS)
Left: Delivered to VF-101 on 28 October 1981, this TARPS Tomcat was assigned to VF-102 in January 1982. It was photographed at NAS Oceana on 29 October 1996 while assigned to VF-32. As of February 1998 it was assigned to VF-14. (Kaston)

F-14A-115-GR, **161286**(TARPS)
Right: This low-vis *Pukin' Dogs* Tomcat was photographed at NAS Oceana on 10 October 1987. Delivered to VF-101 on 30 October 1981, it served with VF-102 before assignment to VF-143 in 1986. On 14 November 1989 it was once again assigned to VF-101 and was stricken while operating near Key West. It departed controlled flight during a 1v1 ACM engagement with a VF-45 Skyhawk. The pilot Ltjg. John T. Burns and RIO Lt. Donovan J. Williams ejected and were recovered without injury. (Author)

F-14A-115-GR, **161287**(B-5)(Upgrade)
Right: The *Evaluators* took delivery of this Tomcat on 24 November 1981. The icon long associated with VX-4 appears on this Tomcat's rudder. During the 1990s the Playboy Bunny logo fell victim to political correctness and has since disappeared from all military aircraft. This Tomcat served with VX-4 until 1987 when it was returned to Grumman Bethpage. It emerged in 1989 as the fifth retrofitted F-14A(Plus). This designation was changed to F-14B on 1 May 1991. Following its upgrade, this Tomcat was once again assigned to VX-4. Additional assignments to VF-101 and VF-102 followed. As of June 1997 it was undergoing UPGRADE modifications at NADEP Jacksonville. (Stewart)

F-14A-115-GR, **161288**
Below: Delivered to VF-1 on 14 November 1981, this Tomcat later saw service with VF-124, VF-1, and VF-211. It was photographed 13 August 1989 while serving with VF-124. As of February 1998 it was in storage at NAS Jacksonville. (Anselmo)

PHOTO NOT AVAILABLE

F-14A-115-GR, **161289**
Left: Delivered to VF-2 on 3 December 1981, this *Bounty Hunters* Tomcat was stricken near Yuma, Arizona on 28 February 1983 when the crew was forced to eject after this Tomcat entered a flat spin.

F-14A-115-GR, **161290**
Left: This VF-1 Tomcat, wearing a faded interim paint scheme, was delivered to the *Wolfpack* on 9 January 1982. It was photographed 15 November 1985. It was stricken on 11 February 1986 while taking part in a night strike escort mission from NAS Fallon. The Tomcat reportedly impacted the ground resulting in the death of both crewmen. (Trombecky/Airframe Images)

F-14A-115-GR, **161291**
Below: This *Bounty Hunters* Tomcat was photographed at NAF Washington during January 1985. Like 161290, it wears the interim gull gray scheme with smaller subdued markings. This Tomcat was delivered to VF-2 on 23 December 1981. It next served with VF-124 and was transferred to VF-211 in 1993. As of February 1998 it was assigned to VF-101. (Author)

F-14A-115-GR, **161292**
Above: Following a 1984 cruise aboard the USS Kitty Hawk (CV-63), the *Wolfpack* of VF-1 and CVW-2 returned to the USS Ranger for a 1986 NorPac/WestPac cruise. This example, also in the interim paint scheme, was photographed during early 1986. Delivered to VF-1 on 7 January 1982, it later would serve with VF-124, VF-21, and VF-51. Recently assigned to VF-101, this Tomcat is currently in residence at Grumman's St. Augustine Florida facility. (Grove)

F-14A-115-GR, **161293**
Right: This VF-111 Tomcat was crewed by "REBEL" and "TAZ" when it was photographed at NAS Fallon early in 1994. Following the *Sundowners'* disestablishment on 31 March 1995, this Tomcat was assigned to VF-101. As of February 1998 it was located at NAS Jacksonville awaiting rework. (Grove)

F-14A-115-GR, **161294**
Right: This VF-1 Tomcat was photographed at NAS Fallon during February 1986, while participating in work-ups with CVW-2. It was delivered to the *Wolfpack* on 28 January 1982. It served with VF-24 and VF-213 before returning to VF-1 in 1992. Following the *Wolfpack's* disestablishment in 1993, it was transferred to VF-111. It served as a Ground Instructional Airframe at NAS Miramar. Following refit it was delivered to VF-32. As of February 1998 it was assigned to VF-211. (Grove)

F-14A-115-GR, **161295**
Left: The *Bounty Hunters* of VF-2 and its sister squadron VF-1 were only assigned to USS Kitty Hawk (CV-63) for one cruise, to the Indian Ocean 13 January 1984 to 1 August 1984. This Tomcat was delivered to VF-2 on 6 February 1982. It later served with VF-124 and VF-111. As of June 1997, this Tomcat was flying with VF-101 and displayed VF-24 commemorative markings. (GB Slides via Jay)

F-14A-115-GR, **161296**
Left: The RIO in this VF-32 Tomcat, photographed 20 October 1995, must be from the *Aardvarks* of VF-114, hence the orange flight suit. It is interesting to note the names of two Captains appear on the canopy rail. One is listed as the CAG, the other, the Deputy CAG. This Tomcat was delivered to VF-1 on 6 March 1982. It later saw service with VF-21, VF-114, VF-2, VF-124, and VF-32. It is currently in storage at NAS Jacksonville. (Author)

F-14A-115-GR, **161297**
Below: When this Tomcat was photographed at Luke AFB in 1985, the *Bounty Hunters* and CVW-2 were assigned to the USS Ranger (CV-61). Of interest is the RIO's helmet which displays the markings of VF-2's sister squadron, VF-1. This example was delivered to VF-2 on 2 March 1982. It later flew with VF-194, VF-114, and VF-213. In February 1998 it was listed in storage at NAS Jacksonville. (Wooley)

F-14A-115-GR, **161298**
Right: This VF-191 Tomcat displays "USS Independence" on its vertical stabilizer. The unit and its sister squadron, VF-194 never made an operational cruise aboard the "Indy." Both squadrons were disestablished less than two years after they stood up. This example was photographed at NAS Willow Grove 14 October 1987. It was stricken on 17 May 1990 while assigned to VF-114. It crashed in restricted area 2301, a TACTS range at 34.00N, 114.03W after departing controlled flight while conducting ACM maneuvers. The aircrew ejected safely. (LeBaron)

F-14A-115-GR, **161299**
Right: Photographed at NAS Miramar on 24 April 1993, this Tomcat was assigned to the *Black Lions* of VF-213. It was delivered to VF-2 on 27 March 1982. From May 1989 to April 1990 it flew with VF-211. It returned to VF-2 from 1990 to 1993. As of February 1998 it could be found at NAS Jacksonville. (Tunney)

F-14A-120-GR, **161416**(B-13)(TARPS)(Upgrade)
Below and opposite top: This VF-143 CAG bird is the 13th of thirty-two F-14As initially upgraded to F-14B standards. The first was delivered to NAS Oceana based squadrons during November 1987. The pod mounted on the front starboard Phoenix pallet is a ALQ-167 ECM pod which is often used in conjunction with TARPS. This Tomcat was delivered to VF-101 on 6 February 1982. It served with VF-143, VF-103, VF-142, and VF-102. (Grove)

See previous entry.

F-14A-120-GR, **161417**(B-8)(TARPS)(Upgrade)
Left: This Tomcat was delivered to VF-101 on 27 April 1982. It served with VF-142 before being returned to the Grumman *"Iron Works"* in December 1987. It emerged as the 8th retrofitted F-14B. It was reassigned to VF-143 and was photographed in their markings near NAS Oceana in 1992. Following service with VF-103 this Tomcat was assigned to VF-102, where it remained assigned through (TARPS). (Ensign Stephen P. Davis USN)

F-14A-120-GR, **161418**(B-4)(Upgrade)
Below: In August 1988 the *Bedevilers* of VF-74 became the first fleet squadron to transition to the F-14B. This VF-74 Tomcat was initially delivered to VF-101 on 21 May 1982; it would later serve with VF-142 before returning to Grumman for modification. It emerged as the 4th F-14B. Since its modification it has been assigned to VF-74, VF-143, VF-142, VF-101, VF-103, and NATC. In December 1996 it was flying with VF-103. As of June 1997 it was awaiting Upgrade modification at NADEP Jacksonville. (Author)

F-14A-120-GR, **161419** (B-9)(Upgrade)
Right: This VF-101 Tomcat was delivered to the *Grim Reapers* on 14 May 1982. It served with VF-142 before becoming the Navy's 9th retrofitted F-14B. It was reassigned to VF-143 and later to VF-101. This example was photographed at NAS Oceana during April 1992. Of interest is the single MK-76 practice bomb loaded on the BRU-42 ITER triple ejector rack. As of February 1998 this Tomcat was assigned to VF-11. (Author)

F-14A-120-GR, **161420**
Below: This VF-101 Tomcat was photographed at the London Ontario International Air Show in June 1982. On 30 August 1983, it was involved in a mid-air collision with 160400 and both Tomcats crashed into the Atlantic Ocean near the Virginia Capes. Two crewmen perished in the crash. (Henderson)

F-14A-120-GR, **161421**(B-17)(Upgrade)
Right: The *Ghostriders* of VF-142, along with its sister squadron, VF-143, transitioned to the F-14B during 1989. This example was photographed at NAS Fallon during April 1991. The squadron and CVW-7 participated in *Operation Desert Shield* and returned to the Persian Gulf during a subsequent cruise from 26 September 1991 to 2 April 1992, aboard the USS Eisenhower (CVN-69). Delivered to VF-101 on 2 June 1982, this Tomcat was assigned to VF-102 as of February 1998. (Grove)

F-14A-120-GR, **161422**(B-18) (TARPS)
Left: This *Ghostriders* CAG bird was photographed at NAS Oceana September 1985. It was delivered to VF-142 on 2 July 1982. It was assigned to VF-101 before returning to Grumman for retrofit as the 18th F-14B. Delivered to VF-103, it served briefly with VF-143 prior to returning to VF-103. As of February 1998 it was listed as being in storage. (Author)

F-14A-120-GR, **161423**
Above: When photographed during October 1982, this VF-142 Tomcat was assigned to the squadron's CO, Cdr. Jack Wood. This Tomcat was stricken on 8 August 1983 while taking part in a Med cruise aboard the USS Eisenhower (CVN-69). (Author)

F-14A-120-GR, **161424**(B-1) (TARPS) (Upgrade)
Left: This VF-142 Tomcat was photographed at NAS Oceana during September 1982. The *Ghostriders* took receipt of it on 2 July 1982. It returned to Grumman early in 1987 and was upgraded to become the first retrofitted F-14B. Following its conversion, it was assigned to NATC, VF-103, and VF-102. As of February 1998 it was assigned to the latter squadron. (Author)

F-14A-120-GR, **161425**(B-19)
Above: This VF-143 Tomcat was photographed while on static display at the 1986 London Ontario International Air Show. It was delivered to the *Pukin' Dogs* on 16 July 1982. Upgraded at Grumman, it became the 19th retrofitted F-14B. This Tomcat was reassigned to VF-143 in 1990. It was stricken on 2 October 1997 while assigned to VF-101 and flying a training sortie. The RIO, Cdr. Craig A. Roll was rescued by helicopter. The pilot, LCdr. Logan A. Allen III, was not recovered. (Author)

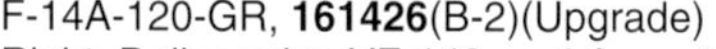

F-14A-120-GR, **161426**(B-2)(Upgrade)
Right: Delivered to VF-143 on 4 August 1982, this Tomcat was still in factory fresh paint when photographed at NAS Oceana during October 1982. Upgraded at Grumman, it became the 2nd retrofitted F-14B, and was returned to the *Pukin' Dogs* in 1989. During January 1997 it was assigned to VF-103. As of February 1998 it was assigned to CFWL, possibly undergoing further modifications. (Author)

F-14A-120-GR, **161427**(B-12)(Upgrade)
Right: The 12th retrofitted F-14B was photographed at NAS Fallon while assigned to VF-142 in April 1991. This Tomcat was originally delivered to the squadron on 10 August 1982. It has been assigned to VF-142 and VF-101. In May 1997, it was being flown by VF-103. As of February 1998 it was undergoing UPGRADE modifications at NADEP Jacksonville. (Grove)

F-14A-120-GR, **161428**(B-6) (TARPS)
Left: Photographed at NAS Oceana in September 1985, this Tomcat was delivered to VF-143 on 2 September 1982. It returned to Grumman in 1987, emerging from the retrofit process as the sixth F-14B, and was assigned to VF-103 in 1989, where it remained through February 1998. (Author)

F-14A-120-GR, **161429**(B-3) (TARPS)(Upgrade)
Left: The Tomcat that would become the third retrofitted F-14B was photographed in VF-142 markings in September 1985. Initially delivered to the *Ghostriders* on 1 September 1982, it was reissued to VF-101 following the conversion process. Following its stay with the Atlantic Fleet Readiness Squadron, it served with VF-143 and VF-103. As of February 1998 it was assigned to VF-102. (Author)

F-14A-120-GR, **161430**(B-22) (TARPS)
Below: This was the 22nd Tomcat to be retrofitted to F-14B standards and was photographed at NAS Oceana during May 1990. On 21 January 1991 it became the eleventh coalition combat loss of the Gulf War. Callsign *Clubleaf* 212, it was assigned to VF-103 and being flown on a TARPS mission by Lt. Devon Jones and Lt. Lawrence Randolph Slade when struck by a modified SA-2 Guideline SAM. Both crewmen successfully abandoned the stricken Tomcat. Lt. Jones was rescued by a USAF MH-53J Pave Low III supported by A-10A Warthogs. Lt. Slade was captured and became a POW. (Author)

F-14A-120-GR, **161431**
Right: Delivered on 2 October 1982, this VF-142 Tomcat was stricken 17 March 1983, following a mid-air collision involving 161439. Both Tomcats crashed approximately 100 miles north of Puerto Rico. (U.S. Navy)

F-14A-120-GR, **161432**(B-24)(Upgrade)
Below: Delivered to VF-143 on 1 October 1982, this Tomcat was the twenty-fourth example modified to F-14B standards. It was photographed at Point Mugu while serving with VX-9 in December 1994. The unusual paint scheme is a holdover from this Tomcat's assignment as a fleet adversary with VF-74. Following service with VF-74 and prior to transfer to VX-9, this Tomcat served briefly with VF-101. As of February 1998 this example was assigned to VF-102. (Roth)

F-14A-120-GR, **161433**(B-7)(Upgrade)
Right: Initially delivered to VF-142 on 2 November 1982, this Tomcat was modified to F-14B standards, becoming the seventh example delivered. Following retrofit, this Tomcat was returned to the *Ghostriders*. It was photographed at NAS Fallon in September 1989. On 13 November 1991, the radome separated while this Tomcat was operating from the USS Eisenhower (CVN-69) in the Persian Gulf. LCdr. Edwards and LCdr. Grundmeier were able to make a successful landing back aboard the "IKE." This Tomcat was flown by VF-103 and as of February 1998 it was undergoing UPGRADE modification by Grumman. (Grove)

F-14A-120-GR, **161434**(B-25)
Left: This F-14B assigned to VF-74 was photographed upon its return from the Gulf War, 27 March 1991. The cruise, which began 7 August 1990, marked the first combat for the F-14B. This Tomcat was delivered to VF-142 on 2 November 1982. Following retrofit, it was assigned to VF-142, VF-74 and VF-143. It was later transferred to VF-32 in 1997 becoming the *Swordsmen's* first F-14B. (Author)

F-14A-120-GR, **161435**(B-26)(Upgrade)
Left: Another happy VF-74 aircrew returns from the Gulf War on 27 March 1991. Corrosion control is a constant battle and resulted in this Tomcat's mottled paint scheme, typical of aircraft which have spent an extended time at sea. This example was delivered to VF-142 on 2 November 1982. Following upgrade as the 26th F-14B, it served with VF-74, VF-101, and NATC. As of February 1998 it was assigned to VF-102. (Author)

F-14A-120-GR, **161436**
Below: With the loss of three Tomcats, 17 October 1983 was not a good day for CVW-7. This example, photographed on its delivery flight, 18 December 1982, was assigned to VF-142 when it suffered a hydraulic failure. The crew was forced to eject over the Caribbean Sea. The other two losses on 17 October 1983 were 161431 and 161439. (Mongeon)

F-14A-120-GR, **161437**(B-15)(Upgrade)
Above: The *Diamondbacks* of VF-102 transitioned to the F-14B in 1994. This example, photographed 23 March 1996, was the 15th retrofitted Tomcat delivered. It was assigned to VF-143 on 25 January 1983 and later served with VF-74, VF-101, and was delivered to VF-102 in October 1994. Following service with VF-11 it was assigned to VF-32 as of February 1998. (JEM Slides)

F-14A-120-GR, **161438**(B-27)
Right: This rather plain VF-101 Tomcat was delivered to VF-143 on 19 January 1983. It served briefly with VF-101 before returning to Grumman. Following retrofit to F-14B standards, it was reassigned to VF-101. (Author)

PHOTO NOT AVAILABLE

F-14A-120-GR, **161439**
Right: Delivered to VF-143 on 24 January 1983, this example was stricken 17 March 1983 following a mid-air collision with 161431.

F-14A-120-GR, **161440**(B-10)
Left: Photographed at NAS Fallon in October 1991, this VF-103 Tomcat was the 10th example modified to F-14B standards. The principal external difference between the F-14A and B are the afterburner cans. Equipped with the more powerful F110-GE-400 powerplants, the F-14B's are significantly larger. This example was delivered to VF-142 on 19 January 1983. During its operational career it has been assigned to VF-103, VF-142, and VF-101. As of February 1998 it was still assigned to the *Grim Reapers* of VF-101. (Grove)

F-14A-120-GR, **161441**(B-16) (TARPS)
Left: During May 1989, the *Pukin' Dogs* of VF-143 began to make the transition to the F-14B. This example was delivered to VF-142 on 3 March 1983. It later served with VF-101, and in 1989 became the first F-14B delivered to VF-143. When the *Pukin' Dogs* embarked aboard the USS Eisenhower (CVN-69), they became the first fleet squadron, along with VF-142, to deploy with the F-14B. This Tomcat was photographed in July 1992 while operating from the USS Dwight D. Eisenhower (CVN-69). This was the 16th Tomcat retrofitted to F-14B configuration. This Tomcat was still assigned to VF-143 as of February 1998. (Lt Jack Liles USN)

F-14A-120-GR, **161442**(B-14)(Upgrade)
Below: On 24 February 1983 this Tomcat was delivered for the first time to VF-143. It was delivered from Grumman a second time in 1990 following its conversion to F-14B standards. Following assignment to VF-102 it was transferred to VF-103 in mid-1998. (Rys)

F-14A-120-GR. **161443**
Right: Photographed at NAS Oceana on 7 August 1985, this VF-2 Tomcat, assigned to the Pacific Fleet, is normally based at NAS Miramar. It was delivered to the *Bounty Hunters* on 31 March 1983 and transferred to VF-111 in 1987. By 1990 it had been delivered to NADEP North Island. It was stricken on 13 January 1992. (JEM Slides)

F-14A-120-GR, **161444**(B-11)
Right: This example was delivered to VX-4 on 26 March 1983. It was modified as the 11th F-14B and reassigned to VX-4 in 1989. It was photographed in formation with a VX-4 F-4J also displaying "Black Bunny" livery. Of interest is the temporary scheme applied to the Tomcat. Later assigned to VF-101 it was stricken on 17 April 1996 when it crashed into a wooded area short of NAS Oceana's runways. The pilot Lt. Ross Slavin and RIO Lt. Dean Kluss ejected safely. (Gann via Kaston)

F-14A-120-GR, **161445**
Below: This *Grim Reapers* Tomcat was photographed at NAS Oceana on 23 March 1996. It was delivered to VF-124 on 13 May 1983. It later served with VF-154, VF-21, VF-51, and VF-101. As of September 1997 it was assigned to VF-101 but was in the process of being stripped of useful parts. (JEM Slides)

F-14A-125-GR, **161597**
Above: According to the markings on this VF-213 Tomcat, it was assigned to CVW-11 aboard the USS Abraham Lincoln when it was photographed at Point Mugu on 16 May 1991. This Tomcat was delivered to VX-4 on 26 April 1983, and later served with VF-21. It was stricken following a mid-air collision with another Tomcat while operating from the USS Abraham Lincoln (CVN-72) on 29 June 1991. (Vasquez)

F-14A-125-GR, **161598**
Left: Photographed on 26 January 1997, this VF-101 Tomcat is marked to commemorate VF-51, which was disestablished on 31 March 1995. This Tomcat was delivered to VF-124 on 2 May 1983. It later served with VF-154 and VF-111. In February 1998 it was still flying with the *Grim Reapers*. (JEM Slides)

F-14A-125-GR, **161599**(B-20)(TARPS)(Upgrade)
Left: This VF-143 Tomcat was photographed at NAS Oceana in May 1994. It was first delivered to VF-124 on 8 June 1983. Retrofitted as the 20th F-14B, it was delivered from Grumman to VF-211 in 1989. It has since served with VF-103 and VF-143. In January 1997 it was assigned to VF-102. On 22 April 1997 it was delivered to AMARC and assigned storage code AN1K0101. By February 1998 it had been returned to service and assigned to VF-32. (Author)

F-14A-125-GR, **161600**
Right: This VF-154 Tomcat was photographed 12 March 1988 at NAS Oceana. It was delivered to VF-124 on 3 June 1983. It served with VF-154 and was transferred to VF-21 in 1991. Following service with VF-211 it was reassigned to VX-9 where it remained through February 1998. (Trombecky/Airframe Images)

F-14A-125-GR, **161601**(B-21) (TARPS)
Below: The *Freelancers* of VF-21 transitioned to the Tomcat from November 1983 to March 1984. This example was photographed at NAS Willow Grove on 13 December 1987. It was delivered to VF-124 on 21 May 1983. It later served with VF-21 prior to being retrofitted as the 21st F-14B. It was delivered to VF-103 and stricken on 13 September 1993. It crashed 40 miles northeast of Cape Hatteras, North Carolina, after the crew was forced to eject. The pilot and RIO were rescued by private fishing boats. (Roop)

F-14A-125-GR, **161602**
Right: The *Black Knights* of VF-154 were paired with VF-21 through most of its Tomcat career. This example was delivered to VF-124 on 20 June 1983. It was photographed 19 May 1987 while assigned to CVW-14 aboard the USS Constellation (CV-64). It was stricken on 24 July 1989. (Van Aken)

F-14A-125-GR, **161603**
Above: This VF-21 CAG bird was photographed at NAS Miramar early in 1991 following CVW-14's first cruise aboard the USS Independence (CV-62). The "Indy" was the first carrier to reach the gulf region in support of *Operation Desert Shield*. This Tomcat was delivered to NATC on 1 July 1983. It later served with VF-124, VF-21, VF-2, VF-24, and VF-213. As of February 1998 it was assigned to VF-14. (Roth)

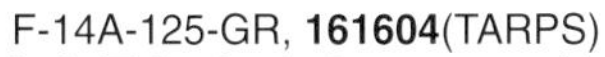

F-14A-125-GR, **161604**(TARPS)
Left: This simple but eye-catching VF-101 CAG bird was photographed at NAS Dallas on 28 November 1987. It was delivered to VF-101 on 12 July 1983. It remained with the *Grim Reapers* until transferred to VF-84 in 1991. On 18 November 1993, it developed an in-flight fire during a ACM training mission. The crew, Lt. Jeffrey Davis and LCdr. Kevin Wensig, were forced to eject. They were picked up by local authorities and a fishing boat. The Tomcat impacted in Currituck Sound, North Carolina. (Wilson)

F-14A-125-GR, **161605**(TARPS)
Left: This VF-32 Tomcat was delivered on 13 July 1983. It was photographed during an Italian port-of-call in January 1995. As part of a deployment which began 20 October 1994, the *Swordsmen* took part in *Operation Southern Watch* over Iraq, and *Operation Deny Flight* conducted from the Adriatic Sea. The tail markings display awards presented the previous year: the "Golden Wrench" for superior maintenance, the Battle "E" for combat readiness, and the Adm. Clifton award as the top fighter squadron in the fleet. (Maglione)

F-14A-125-GR, **161606**
Above: Delivered to VF-124 on 23 July 1983, this Tomcat was later transferred to VF-21. It was stricken on 29 April 1993 while on a training mission off the coast of Japan. Following a suspected engine failure, it departed controlled flight. The pilot, Lt. Kyle Henderson and his RIO, LCdr. Scott Alexander, ejected safely and were rescued. (Author's Collection)

F-14A-125-GR, **161607**
Right: At 0100 hours on 29 August 1995, *Operation Deny Flight* became *Operation Deliberate Force* when armed Tomcats assigned to VF-41 were launched from the deck of the USS Roosevelt (CVN-71). Before the air strikes came to a halt on 1 September, the "Bombcats" had delivered 24,000 pounds of ordnance onto Bosnian Serb targets, with "no escort required." This VF-41 CAG bird displays a small LGB on the nose gear door signifying its role in the first combat delivery of LGBs by a Tomcat. This example was photographed at NAS Oceana 20 October 1995. It was still assigned to the *Black Aces* as of February 1998. (Author)

F-14A-125-GR, **161608**(B-23) (TARPS)
Right: By the time this retrofitted F-14A+ was photographed on the NAS Oceana ramp in September 1995, its designation had been changed to F-14B. This was in large part due to the confusion caused by programming logistics computers with the designations F-14A, F-14A+, and F-14A(Plus). This example was delivered to VF-124 on 27 August 1983 and it also served with VF-21 before being upgraded to F-14B standards. It would later serve with VF-21 and VF-103. As of September 1997, it was assigned to VF-143. (Author)

F-14A-125-GR, **161609**
Above: Thus far in this volume, and at least by 10 June 1986 when this Tomcat was photographed, it would appear the *Freelancers* of VF-21 had yet to familiarize themselves with the low-vis tactical paint scheme. This example was delivered to VF-124 on 6 September 1983. It later served with VF-21 before transfer to VF-114. As of February 1998, it was flying with the Weapons Test Squadron-Point Mugu and was redesignated a NF-14A. (Morgan)

F-14A-125-GR, **161610**(B-30) (TARPS)
Left: The 30th retrofitted F-14B was delivered to VF-143 in January 1994 and photographed at NAS Oceana during May 1995. It was initially delivered to VF-124 as an F-14A on 18 September 1983. Following its retrofit it was assigned to VF-211, VF-103, and to VF-143 where it was serving as of February 1998. (JEM Slides)

F-14A-125-GR, **161611**(TARPS)
Left: The *Fighting Checkmates* of VF-211 transitioned to the F-14A during 1975-76. During 1989 the squadron transitioned to the F-14B, and then back to the F-14A in 1992. This example was photographed at Reese AFB, Texas, during July 1996. It was delivered to VF-101 on 2 November 1983. It later served with VF-31, VF-101, VF-111, and VF-213. As of June 1997 this Tomcat was assigned to VF-211. (McMasters)

F-14A-125-GR, **161612**
Right: The *Gunfighters* of VF-124 took receipt of this Tomcat on 29 October 1983. It was later transferred to VF-154 and photographed in their markings at NAS Miramar in January 1991. When the USS Independence (CV-62) was home ported at NS Yokosuka, Japan, CVW-5 and VF-154 became forward deployed to NAS Astugi, Japan. This example has also served with VF-1 and VF-213. As of February 1998 it was assigned to CFWL. (Puzzullo)

F-14A-125-GR, **161613**
Above: Photographed aboard the USS Constellation (CV-64) on 24 November 1984, this VF-21 Tomcat was stricken on 12 March 1986. It suffered a ramp strike during carrier qualifications aboard the USS Constellation (CV-64). Both aircrew ejected successfully. (Morgan)

F-14A-125-GR, **161614**
Right: This VF-154 Tomcat was photographed at NAS Fallon participating in air wing work-ups for CVW-14's first deployment aboard USS Constellation (CV-64). This example was stricken 20 March 1987. (Grove)

F-14A-125-GR, **161615**
Left: Delivered to VF-124 on 15 December 1983, this Tomcat was later transferred to VF-21 in January 1984. It was photographed 12 March 1988 at NAS Miramar. It later served with VF-111, VF-211, and VF-114. By February 1998 it was assigned to VF-211. (Snyder)

F-14A-125-GR, **161616**
Below: Like its sister squadron VF-21, the *Black Knights* of VF-154 have somehow managed to stall the Navy's fade-to-gray. This example displays high-vis markings applied to the interim overall gull gray scheme. Delivered to VF-124 on 7 December 1983 it has also served with VF-213 and VF-21. As of February 1998 it was serving with VF-101. (Greby)

F-14A-125-GR, **161617**
Left: Parked next to an F-14 from VF-154, its sister squadron, this VF-21 Tomcat was photographed at NAF Washington in December 1986. It was delivered to the *Freelancers* on 21 December 1983. During its operational career it has served with PMTC, VF-111, VF-213, and as of February 1998, was back with VF-154. (Author)

F-14A-125-GR, **161618**

Right: During CVW-14's 1990 cruise aboard the USS Independence (CV-62), each squadron applied special tail markings to their CAG-birds. This example was assigned to VF-154. See 161603 for VF-21's CAG-bird from the same cruise. This Tomcat was delivered to VF-154 on 19 December 1983 and was the first Tomcat received by the *Black Knights*. The squadron's first Tomcat flight took place on 27 December 1983. This example served with VF-51 and VF-21 before returning to VF-154 in November 1995. On 31 August 1997 it was stricken when it crashed 143 miles off the Boso peninsula, Chiba Prefecture, Japan, while practicing take-offs and landings aboard the USS Independence. The aircrew was rescued by the plane-guard helicopter. (Roth)

F-14A-125-GR, **161619**

Right: "Juice" and "Slapshot" were assigned to this VF-21 Tomcat when it was photographed at NAS Miramar during January 1991. This F-14A was transferred to VF-1 and then VF-24. As of February 1998 it was on strength with VF-101. (Puzzullo)

F-14A-125-GR, **161620**(TARPS)

Below: Yes, that is a 2000-lb bomb hanging from this VF-124 CAG-bird. In order to prolong the Tomcat's life, the Navy has dropped its "not a pound for air-to-ground" policy in regards to the Tomcat. This example was photographed at NAS Miramar in March 1992. It was delivered to VF-154 on 9 February 1984. This Tomcat has since served with VF-124, VF-213, and VF-211. As of February 1998 it was assigned to VF-154. (Puzzullo)

F-14A-125-GR, **161621**(TARPS)
Two photos at left: *"Miss Molly"* previously adorned the C-1A COD assigned to the USS Carl Vinson. It was named for Mrs. Molly Snead, who cared for Senator Vinson's wife. When the C-1A was retired, Lt. Mark Conn, an artist, kept the tradition alive on VF-111's CAG bird. This example was photographed at NAF Washington on 2 December 1989. This Tomcat was delivered to VF-21 on 3 February 1984. It later served with VF-111, VF-213, and VF-124. As of February 1998 it was assigned to VF-154. (Handelman)

F-14A-125-GR, **161622**(TARPS)
Below: Delivered to VF-154 on 27 January 1984, this Tomcat was photographed at NAS Fallon in September during air wing work-ups in preparation for a cruise aboard the USS Constellation (CV-64). This example later served with VF-111, prior to returning to VF-154. Following service with VF-213 it was not assigned to another squadron. By February 1998 it was simply assigned to CFWL. (Grove)

F-14A-125-GR, **161623**(D FSD PA-4)
Above: Delivered to NATC on 23 March 1984, this F-14A has remained assigned there for its entire operational career. It was modified by Grumman with a full F-14D avionics suite but retained the TF30-PW-414A powerplants, GRU-7 ejection seats and single chin pod. Designated PA-4, it conducted JTIDS integration and compatibility trials. It was photographed at NATC Patuxent River on 30 November 1992. It was stricken 16 December 1993. (Kaminski)

F-14A-125-GR, **161624**(TARPS)
Right: Delivered to VF-154 on 3 April 1984, this Tomcat was photographed during a 19 May 1987 stopover at NAF Diego Garcia. At the time, CVW-14 was taking part in an Indian Ocean cruise aboard the USS Constellation (CV-64). This example later served with VF-124, VF-2, VF-111, and VF-213. As of February 1998 it was assigned to VF-211. (Van Aken)

F-14A-125-GR, **161625**(TARPS)
Right: This Tomcat was delivered to VF-21 on 5 April 1984. On 2 August 1989 while assigned to VF-111 this Tomcat developed a fire following a catapult launch. The crew attempted to divert to NAS North Island but was forced to eject 6 nautical miles short of the air station. (Trombecky/ Airframe Images)

F-14A-125-GR, **161626**(TARPS)
Above: This official U.S. Navy photo, depicting an F-14A of VF-194, was taken 8 February 1988 prior to the squadron's disestablishment. It was first delivered to VF-154 on 3 April 1984. Following service with VF-194, it was returned to the *Black Knights*. It was assigned to VF-211 and returned to VF-154 for a third tour. Following service with VF-213 it was assigned to CFWL. (LtCdr. Art Legare/USN)

F-14A-130-GR, **161850**
Left: This VF-31 CAG bird was photographed in 1986 during the *Tomcatters* first cruise aboard the USS Forrestal (CV-59) and their first cruise as part of CVW-6. This Tomcat is armed with a pair of AIM-54 Phoenix and AIM-9 Sidewinder missiles. This example was delivered to VF-101 on 25 April 1984. As of February 1998 it was assigned to VF-211. (USN)

F-14A-130-GR, **161851**(B-28)(Upgrade)
Left: Colorful paint schemes are creeping back to Tomcat squadrons as is evidenced by this VF-102 F-14B. It was photographed 2 January 1996, aboard the USS America (CV-66), during its 1995-96 Med/Adriatic cruise in support of *Operation Joint Endeavor*. This would mark the last cruise for the USS America; she was decommissioned upon her return. This Tomcat was delivered to VF-101 on 24 April 1984. It was withdrawn from service for retrofit and emerged as the 28th F-14B. Prior to its transfer to VF-102, it served with VF-101 and VF-142. As of August 1997 it was in storage at AMARC assigned storage code AN1K0102. By February 1998 this Tomcat was again flying, assigned to VF-32. (Author's Collection)

F-14A-130-GR, **161852**
Right: This VF-31 Tomcat is a moment away from touchdown at NAS Oceana. It is armed with a pair of practice CATM-7 Sparrows. The *Tomcatters* somehow managed to keep their colorful schemes much longer than most other squadrons. Delivered to VF-101 on 1 June 1984, it later served with VF-31, VF-41, and VF-84. This F-14A was assigned to VF-32 but as of June 1997 it was resting at Grumman's St. Augustine facility awaiting rework or modification. (Author)

F-14A-130-GR, **161853**
Right: With a lineage reaching back to 1927, the *Red Rippers* of VF-11 are one of the oldest squadrons in the Navy. From Curtiss F6C-3 Hawks to Grumman F-14 Tomcats, the unit has always been known for its colorful markings, These include a Boar's-head insignia reportedly borrowed from a design on bottles of Gordon's gin. This example was photographed at NAF Washington on 12 December 1987. The "AE" tail-code denotes assignment to CVW-6. The other letters appearing on the rudder are awards: the Battle "E" and Safety "S." This example was delivered to VF-11 on 25 May 1984. In March 1996 it was serving with VF-24. As of February 1998 it was serving with VF-211. (Brabant)

F-14A-130-GR, **161854**
Below: Delivered to VF-31 on 8 June 1984, this *Tomcatters* CAG bird was photographed at NAS Oceana in April 1985. On 18 June 1986 this Tomcat crashed when it departed controlled flight while flying over the Mediterranean Sea. One aircrew member lost his life in this mishap. (Author)

F-14A-130-GR, **161855**(B)
Left: U.S. Naval Aviation celebrated its 75th birthday during 1986. There were a number of special paint schemes to commemorate the event including this example from VF-101. This Tomcat was delivered to VF-11 on 19 June 1984. It served with VF-101, VF-11, and VF-14. It was delivered to Grumman's St. Augustine, Florida facility in February 1994 and emerged as an F-14B. As of February 1998 it was again assigned to VF-101. (Author)

F-14A-130-GR, **161856**
Left: The *Tomcatters* bid farewell to the F-14A in 1992, when the squadron relocated to NAS Miramar and transitioned to the F-14D. This F-14A was delivered to VF-31 on 3 July 1984. It later served with VF-24 and VF-213. As of February 1998 it was assigned to VF-211. (Author)

F-14A-130-GR, **161857**
Below: VX-9, Det. Point Mugu was formed following the April 1994 disestablishment of VX-4 and VX-5. This example was photographed during November of 1995. Delivered to VF-101 on 1 August 1984, it served with VF-11 and VX-4 prior to its assignment to VX-9 Det. Point Mugu where it remained through February 1998. (Author's Collection)

F-14A-130-GR, **161858**(B)
Above: This VF-14 Tomcat had just "trapped" aboard the USS John F. Kennedy (CV-67) when it was photographed 13 April 1992. Stationed in the Med, the JFK, with CVW-3 embarked, was taking part in *Operation Provide Promise* providing support for air-drops of food to the population of Bosnia-Herzegovina. This Tomcat was delivered to VF-11 on 17 August 1984. Following service with VF-14 it was delivered to Grumman's St. Augustine facility for retrofit. It emerged as an F-14B and was assigned to VF-101. As of February 1998 it was assigned to VF-143. (Bell)

F-14A-130-GR, **161859**(B)
Right: Delivered to VF-11 on 18 August 1984, this VF-14 Tomcat was later delivered to Grumman's St. Augustine Facility and retrofitted to F-14B standards. It was photographed in VF-14 markings on 11 March 1992 while flying an ACM hop from NAS Fallon during a weapons Det. As of February 1998 it was assigned to VF-101. The VF-14 Tomcat in the background is 161858. (Tunney)

F-14A-130-GR, **161860**(B)(Upgrade)
Right: The black radome indicates this VF-101 Tomcat was formerly assigned to VF-31. It later served with VF-11 and VF-41 prior to being reworked by Grumman at its St. Augustine facility. In January 1997 it was assigned to VF-102, but by June it was flying with VF-11. It was still assigned to the *Red Rippers* as of February 1998. (Author)

F-14A-130-GR, **161861**
Left: The VF-33 *Starfighters* and CVW-1 were taking part in integrated air wing training at NAS Fallon when this F-14A was photographed July 1991. A month later VF-33 embarked aboard the USS America (CV-66) to participate in *Northstar '91*, becoming the first Tomcat squadron to operate in the Frohavet Fjord of Norway. This Tomcat was stricken on 15 December 1992. It crashed while on a training mission 30 miles east of Oregon Inlet, North Carolina. The RIO, Lt. Gregg Hilliard, was rescued by a helicopter from the USS Clark (FFG-11). The pilot, Lt. Joseph Burns, was rescued by a Coast Guard Helicopter from Elizabeth City, NC. (Grove)

F-14A-130-GR, **161862**(B)
Above: This Tomcat was delivered to VF-31 in November 1984. It served with VF-33 and VF-101 before receiving an F-14B upgrade at Grumman's St. Augustine facility in 1994. It later served with VF-74 until that squadron's disestablishment. It was then transferred to VF-101. On 5 September 1997 it was photographed at NAS Oceana displaying this distinctive *Grim Reaper's* paint scheme. (Author)

F-14A-130-GR, **161863**
Left: This VF-14 *Tophatters* Tomcat was photographed at NAS Oceana in June 1996. The "AJ" tail code signifies it is assigned to CVW-8, an Atlantic Fleet Air Wing. This Tomcat was delivered to VF-11 in November 1984. It entered NADEP Norfolk in August 1990 and remained there until 1993. In February 1998, it was assigned to VF-14. (Author)

F-14A-130-GR, **161864**(TARPS)
Right: This VF-31 Tomcat was the 500th built for the Navy. It was delivered in November 1984 and has been assigned to VF-31, VF-101, and VF-32. It was photographed at NAS Oceana in September 1985 wearing VF-31 markings. As of February 1998 it was assigned to VF-41. (Author)

F-14A-130-GR, **161865**(D FSD PA-1)
Right: This Naval Weapons Test Squadron Tomcat is listed as a NF-14D despite retaining its Pratt & Whitney TF-30 powerplants. This example was the first F-14A converted as a Full Scale Development test bed for the F-14D program and made its maiden flight on 23 November 1987. Referred to as PA-1, this Tomcat was used to conduct communications, navigation, radar, and datalink tests. It was stricken on 29 February 1996. In February 1997, the forward fuselage of this Tomcat was cut off for use in bird strike tests. (Vasquez)

F-14A-130-GR, **161866**(TARPS)
Below: This VF-101 F-14A commemorates and keeps alive the tradition of VF-114. The *Aardvarks* were disestablished on 30 April 1993. This Tomcat was delivered 19 December 1984. It served with VF-101, VF-31, VF-102, and VF-41. By February 1998 it was again serving with VF-101. This example was photographed at NAS Oceana 20 October 1995. (Author)

F-14A-130-GR, **161867**(D FSD PA-2)
Left: This Tomcat was the second Full Scale Development F-14D, and was equipped with the F110-GE-400 powerplant. PA-2 made its first flight on 29 April 1988. It was used for TARPS, radar, avionics and environmental tests. Following completion of the test program it was transferred to PMTC and designated an NF-14D. It was transferred between PMTC and VX-4 and later served with NATC before being put on display at the NAWC-AD, Patuxent River Museum. It was officially stricken on 12 April 1995. (Grumman)

F-14A-130-GR, **161868**(TARPS)
Below: During June 1987 the *Tomcatters* of VF-31 were flying from NAS Fallon, taking part in integrated air wing training in preparation for a North Atlantic cruise aboard the USS Forrestal (CV-59). This example was delivered to VF-31 in January 1985. It later served with VF-101, VF-84, and is as of February 1998 was flying with VF-41. (Grove)

F-14A-130-GR, **161869**
Left: "Blue Nose 88" appears on the nose of this VF-11 Tomcat photographed at NAS Oceana on 8 October 1988. These markings make reference to a 1988 cruise to the North Atlantic. Following this cruise aboard the USS Forrestal (CV-59), the *Red Rippers* and CVW-6 were awarded a Meritorious Unit Commendation. This F-14A was delivered to VF-11 in January 1985. During its service life it flew with VF-24 and VF-124. Later assigned to VF-213, as of February 1998 it was listed as being in storage and assigned to CFWL. (Miller via Sheets)

F-14A-130-GR, **161870**(B-31)(Upgrade)
Above: Delivered in February 1985, this Tomcat was later upgraded to F-14B standards and has served with VF-74, VF-143, and VF-101. In January 1997 it was serving with VF-102. In February 1998 it was resting at NAS Jacksonville. This Tomcat was photographed on Oceana's wash rack in May 1994 while assigned to VF-143. (Author)

F-14A-130-GR, **161871**(B-29)(Upgrade)
Right: Photographed at NAS Oceana in October 1987, this VF-11 Tomcat became the twenty-ninth F-14A to be upgraded to B standards. It was reissued to VF-101 and later served with VF-103. As of February 1998 it was assigned to VF-102. (Author)

F-14A-130-GR, **161872**
Right: This *Red Rippers* CAG bird was photographed at NAF Washington in December 1986. It was delivered to VF-11 in March 1985. On 12 August 1987 it crashed into the North Atlantic following an unsuccessful aerial refueling attempt. Both aircrew were rescued and treated for minor injuries. (Author)

F-14A-130-GR, **161873**(B-32)
Left: Delivered to VF-11 in March 1985, this *Red Rippers* Tomcat was photographed at Nellis AFB on 2 April 1987. It was later upgraded to F-14B standards and returned to the fleet. On 10 June 1996, this Tomcat was serving with VF-103. It made history by dropping and successfully guiding a PAVE Way III GBU-24B/B, laser-guided bomb. The crew for this mission was Lt. Pete Hooper of VX-9 and LCdr. Pete Matthews of VF-103. The bomb scored a direct hit on a target at NAWC China Lake. As of February 1998 this F-14B was still assigned to VF-103. (Rogers)

F-14A-135-GR, **162588**
Left: The *Gunfighters* of VF-124 trained Tomcat crews for more than two decades, 1973 to 1994. This example was photographed at NAF El Centro in April 1987. Small detachments of fighter and attack aircraft deploy to El Centro for air-to-ground operations and for air-to-air gunnery practice against towed targets. This Tomcat later served with VF-213 and VF-24. As of June 1997 it was flying with VF-211. (Roth)

F-14A-135-GR, **162589**
Below: Photographed at Point Mugu in November 1987, aside from the 451 side number, this VF-124 Tomcat is marked identical to 162588. It was delivered in November 1985 and has since served with VF-154. As of February 1998 it was in storage at NAS Jacksonville. (Roth)

F-14A-135-GR, **162590**

Right: Photographed at Miramar in July 1988, this rather vanilla VF-24 Tomcat was delivered in November 1985. It was stricken on 25 January 1993 while serving with the *Renegades*. This Tomcat suffered a catastrophic fire while in FCLP pattern at NAS Miramar. The crew ejected successfully and the aircraft crashed into a landfill next to the base. (Romano)

F-14A-135-GR, **162591**

Two below: This eye-catching VF-124 paint scheme deserves two images to fully appreciate it. The left side was photographed in July 1989 and the right side in January 1990. This F-14A later served with VF-51 and VF-101. In February 1998 it was assigned to VX-9. (Anselmo)

F-14A-135-GR, **162592**
Left: Hauling a pair of dumb bombs, this VF-51 F-14A taxies out for a practice air-to-ground sortie at NAS Fallon in March 1994. A year later, the *Screaming Eagles* would be disestablished, another victim of the post cold-war meltdown (I mean drawdown). During its service life, this Tomcat has flown with VF-1, VF-51, and VF-21. As of February 1998 it was flying with VF-154. (Grove)

PHOTO NOT AVAILABLE

F-14A-135-GR, **162593**
Left: This Tomcat was delivered in January 1986 and was stricken on 17 August 1987 when it departed controlled flight and crashed into the North Pacific Ocean. The aircrew ejected and were rescued.

F-14A-135-GR, **162594**
Below: Delivered in January 1986, this F-14A served with VF-124, VF-111, VF-51, and VF-21. While serving with VF-111, it was affectionately known as the "Buick" or the "Jeepney." The sharkmouth and sunburst are recognized around the world as belonging to VF-111, the *Sundowners*. The "Buick" made its final flight while assigned to VF-111 on 28 September 1992. It entered SDLM for rework and was assigned to VF-154. (Grove)

F-14A-135-GR, **162595**(D FSD PA-3)
Right:This F-14A was delivered 30 October 1985 and was modified by Grumman as the third Full Scale Development F-14D. Designated PA-3, it was utilized to perform ECM, sensor and weapons integration tests. Photographed at NATC in December 1992, it was stricken 28 May 1993 and is now displayed at NAWS Patuxent River. (Kaminski)

F-14A-135-GR, **162596**
Below: Delivered to VF-124 on 31 October 1986, this VF-2 Tomcat was photographed at NAS Miramar in March 1988. Markings consist of a Battle "E" for combat readiness, and "NE" tail code signifying assignment to CVW-2. The red, white and blue stripe is referred to as the "Langley Stripe" and is a carryover of markings displayed when Fighter Squadron 2 was deployed aboard the USS Langley, the Navy's first aircraft carrier. This Tomcat was stricken on 22 September 1988 when it departed controlled flight during an ACM training mission. The aircrew successfully ejected. (Roth)

F-14A-135-GR, **162597**
Right: A mix of Sparrow and Sidewinder missiles are used for air superiority and escort missions. The Phoenix missile is favored for fleet defense. A mix of all three can be carried as depicted on this Tomcat. This example was delivered to VF-1 in October 1985. It was assigned to USS Ranger (CV-61) with CVW-2. It later served with VF-111 and as of February 1998 it was flying with VF-154. (USN)

F-14A-135-GR, **162598**
Above: This VF-2 Tomcat was photographed while flying from the USS Ranger (CV-61) while taking part in *Operation Desert Storm.* Missions included TARPS, BarCAP, MiGSweep, and MiGCAP for CVW-2 Intruders, Italian Tornados, and USAF B-52s. The squadron was also credited with strafing three Iraqi naval vessels, forcing two to go aground on the coast of Kuwait. Delivered in November 1985, it also served with VF-114 and VF-124. As of June 1997 it was assigned to VF-211. (VF-2 via Kaston)

F-14A-135-GR, **162599**
Left: This Tomcat was delivered to VF-1 in November 1985. It was transferred to VF-213 in 1993, and was stricken while assigned to that squadron. On 29 January 1996, at approximately 0945 hours, this Tomcat made an unrestricted climb from Berry Field ANG Facility, Nashville International Airport. The Tomcat departed controlled flight and crashed. The crew, pilot LtCdr. Stacy Bates and RIO Lt. Graham Alden Higgins perished along with three civilians on the ground. This example was photographed at NAS Miramar on 24 April 1993. (Kaston)

F-14A-135-GR, **162600**
Left: Delivered to VF-2 in November 1985, this *Bounty Hunters* Tomcat was photographed 9 February 1990 in a temporary water-based paint scheme applied at NAF El Centro. This Tomcat was transferred to VF-211 in 1992 and was flying with that squadron as of February 1998. (Snyder)

F-14A-135-GR, **162601**

Right: This Tomcat was photographed shortly before VF-1 and its sister squadron, VF-2, was embarked aboard the USS Ranger (CV-61) for a 1992 Indian Ocean/Persian Gulf cruise. The *Wolfpack* and *Bounty Hunters* provided combat air patrols in support of *Operation Southern Watch.* This marked the final cruise for VF-1 and the Ranger. The squadron was disestablished on 1 October 1993. This Tomcat was transferred to VF-114 and later to VF-211. The Ranger was decommissioned 10 July 1993. On 11 March 1993, a Tomcat from VF-124 made the final landing aboard this historic ship. It was Ranger's 330,683rd and last arrested landed. As of February 1998 this Tomcat was assigned to VF-154. (Author)

F-14A-135-GR, **162602**

Right: This VF-51 CAG bird was photographed on 29 July 1990 following the *Screaming Eagles* final cruise aboard the USS Carl Vinson (CVN-70). This Tomcat was stricken 11 July 1994 when it broke in half upon landing aboard the USS Kitty Hawk (CV-63). The cockpit section slid down the deck and over the side. The crew was able to eject but the pilot was severely injured after landing in burning fuel on the flight deck. The NFO was unhurt. Ten deck personnel received minor injuries. (Anselmo)

F-14A-135-GR, **162603**

Below and opposite top: As potent as the Tomcat's arsenal is, the type only scored one aerial victory during the Gulf War. On 6 February 1991, Lt. Stuart "Meat" Broce and his RIO, Cdr. Ron "Bongo" McElraft were airborne in this F-14A. It was armed with four AIM-7M Sparrows and four AIM-9M Sidewinders. In addition, it was carrying a pair of 267 gallon FPU-1 external fuel tanks. Vectored north into Iraq, they were directed to a "bandit" by a USAF E-3 "Sentry." They visually acquired an Iraqi Mi-8 "Hip" helicopter skimming across the desert. When the range between the Tomcat and Mi-8 had closed to within a mile, Lt. Broce launched a Sidewinder which found and destroyed its target. This VF-1 Tomcat was photographed at NAS Miramar on 7 June 1991. It was transferred to VF-124 and then to VF-41. By February 1998 it was assigned to CFWL. (Huston)

See previous entry.

F-14A-135-GR, **162604**
Left: When photographed at NAS Fallon in March 1994, VF-51 was preparing for its last cruise, which, due to mounting tensions in Korea, saw the *Screaming Eagles* and CVW-15 head to the Western Pacific aboard the USS Kitty Hawk (CV-63). The squadron returned to NAS Miramar on 22 December 1994. On 31 March 1995, VF-51 was disestablished. This example later flew with VF-21 and VF-101. In September 1997 it was flying with VF-14. By February 1998 it was assigned to CFWL. (Grove)

F-14A-135-GR, **162605**
Below: This VF-124 Tomcat was delivered in January 1986. Later transferred to VF-1, it was stricken on 15 July 1987 when it departed controlled flight and crashed into the North Pacific Ocean. The aircrew ejected successfully suffering only minor injuries. (McGarry)

F-14A-135-GR, **162606**
Right: This VF-2 Tomcat taxis out for a FFARP mission at NAF El Centro in February 1990. Compare this crude hand-painted temporary scheme to the standard *Bounty Hunter* markings as applied to 162598. This Tomcat was delivered to VF-2 in December 1985. It was transferred to VF-213 in 1992. As of February 1998 it was assigned to VF-211. (Carlson)

F-14A-135-GR, **162607**
Above: The *Wolfpack's* next to last cruise, December 1990 to June 1991, would be an historic one, to the Persian Gulf in support of *Operation Desert Shield* and *Desert Storm.* Embarked aboard the USS Ranger (CV-61) was CVW-2, the only "All-Grumman" air wing. Also referred to as the "Iron Works" air wing, it was made up of Grumman Tomcats, Hawkeyes, Intruders, and Prowlers. During the Gulf War, VF-1 flew more than 540 combat sorties. This example, photographed July 1990 at NAS Miramar, was transferred to VF-24 in 1992. It later served with VF-124 VF-101, and VF-32. As of February 1998 it was assigned to VX-9. (Roth)

F-14A-135-GR, **162608**
Right: Delivered to VF-1 in February 1986, this Tomcat later served with VF-2, VF-11, and VF-24. It was photographed during August of 1996 while assigned to VF-211, and as of February 1998 remained assigned to the *Flying Checkmates.* (Greby)

F-14A-135-GR, **162609**
Left: This Tomcat was delivered to VF-124 in February 1986. It was photographed at Scott AFB during July 1986. Later transferred to VF-1, this Tomcat crashed into the Gulf of Oman on 8 September 1988 when it departed controlled flight while taking part in a ACM training flight. (Van Aken)

F-14A-135-GR, **162610**
Left: Initially delivered to VF-1 in 1987, this Tomcat was photographed in VF-51 markings at NAS Fallon during April 1992. The squadron was working up for a cruise which would see the *Screaming Eagles* flying missions over Somalia in support of *Operation Restore Hope.* On 13 January 1993, the squadron participated in air strikes against Iraq as part of *Operation Southern Watch.* This Tomcat was transferred to VF-213 following the disestablishment of VF-51 on 31 March 1995. As of February 1998 it was assigned to VF-154. (Grove)

F-14A-135-GR, **162611**
Below: This Tomcat was delivered on 25 March 1986. During its service life it served with VF-1, VF-111, and VF-213. This example was photographed in October 1991 lifting off from NAS Miramar. In May 1997 it was assigned to VF-213. By September it was listed on the roster of VF-154 aboard the USS Independence (CV-62). (Author)

F-14A-140-GR, **162688**
Above: Following the Gulf War, VF-84 and CVW-8 took part in *Operation Provide Comfort*. This *Jolly Rogers* CAG bird was photographed on 26 June 1991, the day VF-84 and its sister squadron VF-41 returned to NAS Oceana following the Gulf War. As of February 1998 this Tomcat was assigned to VF-14. (Author)

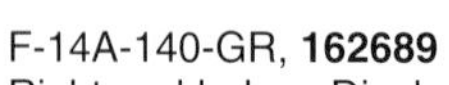

F-14A-140-GR, **162689**
Right and below: Displaying "Bombcat" nose art denoting the Tomcat's new air-to-ground mission, this F-14A was assigned to VF-41 when it was photographed at NAS Oceana in September 1993. It was delivered in April 1986, and has served with VF-41 and VF-101. Following the Gulf War this Tomcat sported nose art and the name "Queen of Spades." In February 1998 it was flying with VF-14. (Author)

F-14A-140-GR, **162690**
Above: Delivered to VF-101 in May 1986, this Tomcat was later transferred to VF-32. At 2245 hours on 6 October 1989 it was stricken after impacting the port jet blast deflector onboard the USS John F. Kennedy (CV-67). The pilot, Lt. Russell C. Walker and his RIO, Lt. Robert Schrader, ejected and were rescued by an SH-3 from HS-7. This Tomcat was photographed 31 January 1989 upon its return from a Med cruise involving the downing of a pair of Libyan MiG-23s. (Author)

F-14A-140-GR, **162691**(B)(Upgrade)
Left: This VF-14 Tomcat wears a temporary water-based paint scheme during a 1989 deployment to NAS Fallon. These paint schemes were reportedly easy to apply with brushes, rollers, even brooms, but were a "bitch" to remove. This Tomcat was delivered in May 1986. It served with VF-14, VF-33, and VF-102. As part of a purchase of additional retrofits, this F-14A was delivered to Grumman's St. Augustine Facility and upgraded to F-14B standards. In 1995 it was assigned to VF-103. It was one of the first F-14Bs received by VF-11 following the *Red Rippers'* transition from the F-14D to the F-14B which commenced during June 1997. As of February 1998 it was assigned to VF-32. (Grove)

F-14A-140-GR, **162692**(B)(Upgrade)
Left: Delivered to VF-101 in June 1986, this Tomcat was later transferred to VF-143 and then to VF-84 in 1990. It was photographed at NAS Oceana on 26 June 1991 upon the *Jolly Rogers'* return from *Operation Desert Storm.* In 1994 this example was delivered to Grumman's St. Augustine Facility and retrofitted as an F-14B. In 1995 it was assigned to VF-103. Following service with VF-102 it was flying with VF-11 as of February 1998. (Author)

F-14A-140-GR, **162693**(B)
Above: This Tomcat was delivered to VF-74 in June 1986. It served with VF-33 and VF-101 before being delivered to Grumman's St. Augustine Facility where it was retrofitted to F-14B standards. As of February 1998 it was assigned to VF-101. (Trombecky/Airframe Images)

F-14A-140-GR, **162694**(B)(TARPS)
Right: The only kill markings recorded following VF-32's engagement with a pair of Libyan MiG-23s on 4 January 1989, were carried on this *Swordsmen* boss bird. They consist of two small Libyan flags painted on the canopy rails. These markings quickly disappeared following the squadron's return to NAS Oceana. This example was photographed 31 January 1989 during the squadron's homecoming. Delivered in July 1986, it served with VF-32 and VF-101 prior to being modified to F-14B standards at Grumman's St. Augustine Facility. As of February 1998 it was assigned to VF-103. (Author)

F-14A-140-GR, **162695**(B)(TARPS)
Right: This low-vis VF-102 CAG bird was delivered in August 1986. Its service with the *Diamondbacks* included several combat sorties during the Gulf War. In 1994-95 it was retrofitted to F-14B standards at Grumman's St. Augustine Facility and delivered to VF-143. It was still assigned to VF-32 as of February 1998. (Author)

F-14A-140-GR, **162696**
Above: This *Sluggers* Tomcat was photographed in April 1989, during VF-103s final year of operating the F-14A. Teamed with VF-74, both squadrons converted to the much improved F-14B later in the year. This VF-103 CAG bird was delivered in July 1986. It served with VF-103, VF-102, VF-101, and VF-32. As of February 1998, it was assigned to VF-41. (Author)

F-14A-140-GR, **162697**
Left: For over a decade, the Navy has hosted exchange crews from our allies. This VF-101 F-14A was the personal mount of Squadron Leaders Rich Powell and John Carter, RAF. This Tomcat was delivered in August 1986. Following service with VF-32 it was transferred to VF-41 where it was still assigned as of February 1998. (Author)

F-14A-140-GR, **162698**
Left: This VF-32 Tomcat was photographed aboard the USS Eisenhower (CVN-69) in January 1995, during an Italian port-of-call. The embarked air wing CVW-3 was making its first cruise aboard the "IKE." This F-14A was delivered in September 1986. It served with VF-33 and flew a number of combat sorties during the Gulf War. As of February 1998 it was assigned to VF-14. (Maglione)

F-14A-140-GR, **162699**(B)(Upgrade)
Right: Delivered in October 1986, this VF-101 Tomcat was photographed at NAS Oceana on 13 April 1992. In 1994 it was delivered to Grumman's St. Augustine Facility for retrofit. It emerged as an F-14B and was reassigned to VF-101. As of February 1998 it was assigned to VF-11. (Author)

F-14A-140-GR, **162700**(B)
Right: This VF-14 CAG bird, with its half-blue/half-gray paint scheme, is the result of corrosion control efforts, not a new or experimental camouflage. It was photographed 27 March 1991, returning to NAS Oceana following the Gulf War. This Tomcat was delivered in October 1986 and served with VF-14 and VF-102. It was upgraded to F-14B standards in 1994-95 at Grumman's St. Augustine facility. As of February 1998 it was assigned to VF-103. (Author)

F-14A-140-GR, **162701**(B)(TARPS)
Below: No U.S. Navy F-14 has recorded two official aerial victories as depicted on this VF-32 Tomcat. These kill markings are representative of the squadron's scoreboard, specifically, from an engagement with two Libyan MiG-23s on 9 January 1989. This F-14A was photographed at NAS Oceana in May 1990. Delivered in October 1986, it served with VF-32. Following modification to F-14B standards it was delivered to VF-103. (Author)

F-14A-140-GR, **162702**
Left: Photographed on 6 December 1989, this VF-84 Tomcat made the first arrested landing aboard the USS Abraham Lincoln (CVN-72) on 1 December 1989. It was piloted by "Abe's" Captain, William B. Hayden, and Cdr. Charles K. Crandall Jr., CVN-72 ops officer. This example was stricken on 5 June 1990. (PH3 Gary Ward))

F-14A-140-GR, **162703**(B)(TARPS)
Left: The *Black Aces* of VF-41, along with its sister squadron VF-84, were the last Tomcat squadrons to return to NAS Oceana following *Operation Desert Storm.* This VF-41 CAG bird was photographed during its homecoming, 26 June 1991. This Tomcat was delivered in December 1986. Following service with VF-41 it was delivered to Grumman's St. Augustine Facility for retrofit to F-14B standards in 1994. It was assigned to VF-143 in November 1995. On 12 August 1997 it was involved in an unusual deck accident. The pilot was ejected while landing aboard the USS Stennis (CVN-74). The RIO was rescued as the Tomcat sat on the flight deck with its engines running. The pilot was rescued immediately by a plane guard helicopter from HS-5. (Author)

F-14A-140-GR, **162704**
Left: On 29 August 1995, VF-41 became the first Tomcat squadron to deliver air-to-ground ordnance in combat. This event occurred on the first day of *Operation Deliberate Force.* In addition to the delivery of fifteen tons of laser guided bombs, the *Black Aces* also launched thirty-nine ADM-141 Tactical Air Launched Decoys (TALD). This example was photographed in August 1995 aboard the USS Roosevelt (CVN-71). Delivered 11 December 1986 it flew with VF-102 before its transfer to VF-41. It was assigned to the *Black Aces* as of February 1998. (Zambon)

F-14A-140-GR, **162705**(B)(TARPS)
Right: Anticipating war in Southwest Asia, the *Starfighters* of VF-33 deployed ahead of schedule, arriving in the Red Sea on 16 January 1991. They flew their first combat mission on 19 January. It consisted of a day strike against the Latifiya Scud missile facility. This example was photographed 17 April 1991 during VF-33's homecoming at NAS Oceana. It was later retrofitted to F-14B standards by Grumman's St. Augustine Facility. As of February 1998 it was assigned to VF-143. (Author)

F-14A-140-GR, **162706**
Right: Delivered in January 1987, this VF-103 Tomcat was photographed 28 May 1988 while assigned to the USS Independence (CV-62). It was stricken on 18 April 1989 while assigned to VF-101. Piloted by Ltjg. Thomas A. Cooper with Lt. Christopher C. Cinnamon as instructor RIO, the Tomcat entered a stall while on route to restricted area W-174C near NAS Key West. Lt. Cinnamon initiated ejection prior to impact with the water. Both aircrew survived without injury. (Rys)

F-14A-140-GR, **162707**
Below: On 22 September 1987, this VF-74 Tomcat, crewed by Lt. Tim Dorsey and LtCdr. Ed Holland, added a Phantom II to the F-14's scoreboard of aerial victories. The victim, a 26th TRW RF-4C, serial 69-0381, crewed by Lt. Randy Spouse and Capt. Michael Ross, was taking part in Display Determination involving aircraft from the USAF and USS Saratoga (CV-60). It was downed by a AIM-9 Sidewinder. This Tomcat was delivered in January 1987. It served with VF-74 and VF-101. It was stricken, on 17 January 1993. (Author)

F-14A-140-GR, **162708**
Above: The *Hunters* of VF-201 and the *Superheats* of VF-202 received a mix of older F-14A's rebuilt to Block 135 standards and the last new-built F-14As. This example was delivered in February 1987. It was photographed landing at NAS Dallas on 21 February 1988 while assigned to VF-201. It was stricken on 23 December 1992. LCdr. "Jim Bob" Segars lost his life in the crash. (Snyder)

F-14A-140-GR, **162709**
Left: This VF-201 CAG bird was delivered in February 1987. It served with the *Hunters* until 1995 when it was transferred to VF-101. As of September 1997 it was listed as resting at NADEP Jacksonville. (Leekey)

F-14A-140-GR, **162710**
Left: Delivered in March 1987, this VF-202 Tomcat was photographed at Bergstrom AFB during the 1988 Reconnaissance Air Meet. Up against dedicated recce platforms such as the RF-4C, the TARPS-equipped Tomcats did not fare well. When VF-202 was disestablished in 1995, this example was transferred to VF-101, where it was assigned as of June 1997. (Author)

F-14A-140-GR, **162711**
Right: The last F-14A was delivered to the fleet on 31 March 1987. It was photographed while assigned to VF-202. With the delivery of TARPS, the *Superheats* became CVWR-20's dedicated photo reconnaissance asset. On 9 July 1994, VF-202 was disestablished and this Tomcat was transferred to VF-201. It later served with VF-101 prior to being dispatched to NADEP Jacksonville. (Grumman via Kaston)

F-14B-145-GR, **162910**
Right: The first new-built F-14B was delivered to NATC in November 1987. The major difference between the F-14A and B involves the installation of F110-GE-100 turbofans providing a significant increase in thrust; so much so that the F-14B can launch from a carrier in military power without having to resort to fuel-guzzling afterburners. Other changes include the deletion of the glove vanes and addition of the AN/ARC-182 radio and AN/ALR-67 Radar Warning system. This example was photographed September 1995 while assigned to VF-102. It is currently assigned to CFWL. (Author)

F-14B-145-GR, **162911**(Upgrade)
Below: The *Fighting Renegades* of VF-24 transitioned to the F-14B in 1989. Due to an insufficient number of F-14Bs, the squadron transitioned back to the F-14A in 1992. This example was photographed 14 August 1989 while assigned to VF-24. It was delivered in December 1987 and has served with VF-24, Topgun, and VX-9. As of February 1998 it was assigned to VF-11. (Anselmo)

F-14B-145-GR, **162912**(Upgrade)
Above: Delivered in February 1988, this F-14B has served with VF-24, VF-101, VF-142, VF-143, and VF-102. As of February 1998 it was assigned to VF-11. This example was photographed 14 August 1989 while assigned to VF-24. Of intrest are the false canopy markings. (Anselmo)

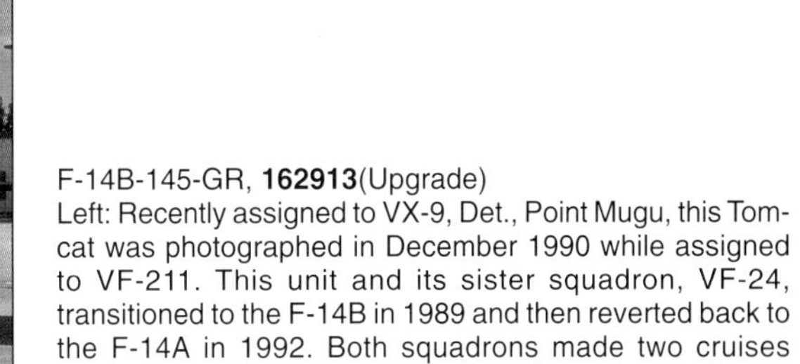

F-14B-145-GR, **162913**(Upgrade)
Left: Recently assigned to VX-9, Det., Point Mugu, this Tomcat was photographed in December 1990 while assigned to VF-211. This unit and its sister squadron, VF-24, transitioned to the F-14B in 1989 and then reverted back to the F-14A in 1992. Both squadrons made two cruises aboard the USS Nimitz (CVN-68) while equipped with the F-14B. As of February 1998 this example was flying with VF-103. (Roth)

F-14B-145-GR, **162914**
Left: Delivered to VF-101 in March 1988, this F-14B later served with VF-74 and VF-142. It was photographed in VF-142 markings at NAS Oceana on 23 January 1993. In November 1993 it was transferred to VF-103. It was stricken 12 February 1994 while on a Med cruise assigned to CVW-17, embarked aboard the USS Saratoga (CV-60). During ACM training near the Italian coast, it collided with a CVW-17 Hornet. The Tomcat crew ejected and were rescued by a helicopter dispatched from the USS Vicksburg. The damaged Hornet landed safely at Brindisi, Italy. (Rogers)

F-14B-145-GR, **162915**
Above: Delivered in April 1988, this VF-142 *Ghostriders* Tomcat was photographed at NAS Oceana 13 April 1992, following the squadron's final cruise aboard the USS Eisenhower (CVN-69). As of February 1998, it was assigned to VF-143. (Author)

F-14B-145-GR, **162916**(Upgrade)
Right: Delivered in April 1988, this F-14B served with VF-143 and VF-103. It was photographed at NAS Oceana in VF-102 CAG markings on 26 April 1997. It was still flying with the *Diamondbacks* through February 1998. (Author)

F-14B-145-GR, **162917**(Upgrade)
Right: The *Pukin' Dogs* of VF-143, along with its sister squadron, VF-142, were the first two Tomcat squadrons to deploy aboard the USS George Washington (CVN-73). This example, in VF-143 markings, was photographed 30 July 1993. Later transferred to VF-103, it was assigned to VF-102 as of February 1998. (Greby)

F-14B-145-GR, **162918**
Above: Delivered in July 1988, this F-14B was initially assigned to VF-102. It was photographed 2 December 1994 at NAS Oceana. With the decommissioning of the USS America (CV-66) on 9 August 1996, the *Diamondbacks* ended a decade long association with this country's namesake. As of February 1998 this Tomcat was assigned to VF-101. (Author)

F-14B-145-GR, **162919**(Upgrade)

Left: This F-14B was delivered to VF-74 on 11 August 1988. Following service with the *Bedevilers*, it was delivered to NAWC-AD on 11 April 1994, where it underwent Integration Testing of its EW suite. Tests concentrated on integration of installed ALR-67 and ALQ-126B with the ALQ-167 external pod. Following completion of these tests it was delivered to NADEP on 1 June 1994. Later assigned to VF-102, in December 1996 it was at Grumman's St. Augustine facility where it was undergoing conversion to F-14B(Upgrade) and receiving TARPS modifications. As of February 1998 it was assigned to VF-11. (USN)

F-14B-145-GR, **162920**(Upgrade)
Left: Following an eight month deployment during *Operation Desert Shield/Desert Storm*, the *Sluggers* of VF-103 deployed to NAS Fallon for the Advanced Strike Syllabus. This example was photographed at NAS Fallon in October 1991. The mottled paint scheme, consisting of three shades of gray, is interesting. In December 1991, VF-103 reached a twelve year milestone for Class "A" mishap-free operations. In 1990, the squadron was awarded the CNO's Safety "S", which is proudly displayed on the tail of this Tomcat. As of February 1998 this Tomcat was assigned to VF-102. (Grove)

F-14B-145-GR, **162921**(Upgrade)
Right: This VF-103 CAG bird was photographed at NAS Fallon in October 1991, while participating in the Advanced Strike Syllabus. Compare these markings with 162696, 163215, and 163216 to view the evolution in VF-103 markings from 1989 to 1995. This Tomcat was assigned to VF-143 as of February 1998. (Grove)

F-14B-145-GR, **162922**(Upgrade)
Right: This F-14B served with the *Bedelivers* of VF-74 and saw combat during its first cruise. It was transferred to VF-142 and later delivered to Grumman's St. Augustine facility for conversion to an F-14B(Upgrade). As of February 1998 it was serving with the *Diamondbacks* of VF-102. It was photographed 24 April 1997 at NAS Oceana. (Author)

F-14B-145-GR, **162923**
Below: This Boss bird assigned to VF-101 was photographed on 20 October 1995. The eye-catching paint scheme was short-lived. This Tomcat was delivered in September 1988. It served with VF-74, VF-142, and VF-101. As of February 1998 it was assigned to VF-143. (Author)

F-14B-145-GR, **162924**
Left and below: Photographed May 1996, being "fed to the cat", this F-14B was on Alert-5 status aboard the USS Enterprise (CVN-65). Don't let the *Jolly Roger* fool you, this Tomcat is assigned to VF-103. The Enterprise and CVW-17 were part of a joint service exercise including elements of the Royal Navy, Air Force, and Army. Of interest is the AN/AAS-14 LANTIRN pod which is a modified version of the targeting pod used by USAF F-15Es and late-block F-16s. The AN/AAQ-13 navigation pod was omitted due to the extensive software modifications required to carry it on the Tomcat. The *Jolly Rogers* were the first Tomcat squadron to deploy with the laser targeting pods. (Author)

F-14B-145-GR, **162925**
Left and above: In August 1995, the USS America (CV-66) was rushed to the Adriatic Sea on what would be her last cruise. The embarked air wing CVW-1 took part in air strikes against selected Bosnian Serb targets. The crew of this VF-102 "Bombcat", hauling a pair of iron bombs, keep their hands clear of all cockpit switches as the roof-crew prepare it for launch. On 23 February 1996, this Tomcat had the distinction of making the 346,843rd and final launch from the USS America. As of February 1998 this Tomcat was assigned to VF-101. (Maglione)

F-14B-145-GR, **162926**
Left: This VF-103 Tomcat was photographed at NAS Miramar on 19 October 1989. At the time, the squadron was evaluating the F-14B's ALR-67 Radar Warning Receiver (RWR) in Topgun's power projection course. This example was delivered to VF-103 in November 1988. It later served with VF-142 and VF-143. By February 1998 it was flying with VF-32. (Anselmo)

F-14B-145-GR, **162927**
Above: NAS Oceana, the venue for this shot, is a very busy Naval Air Station, especially now that it is host to all but two of the Navy's operational Tomcat squadrons. Delivered to VF-103 in November 1988, this F-14B was assigned to VF-101 before reaching the *Pukin' Dogs* of VF-143. It was photographed 20 October 1995 preparing to launch from NAS Oceana. By February 1998 this Tomcat was assigned to VF-32. (Author)

F-14B 150-GR, **163215**
Right: With the introduction of the "Bombcat", sights like this are common on the Oceana ramp. This bomb trailer is loaded with inert high drag BDU-45s with BSU-86 high drag fin kit. The VF-103 F-14B in the background was delivered in December 1988. It was photographed in September 1995 and as of February 1998 it was still serving with VF-32.

F-14B-150-GR, **163216**
Below right and below: At first glance, this appears to be a VF-84 F-14B landing at NAS Oceana on 20 October 1995. This Tomcat is actually assigned to VF-103. With the disestablishment of VF-84 on 1 October 1995, the "Bones" were passed to VF-103. The "Bones" are those of Ens. Jack Ernie, who was assigned to the *Jolly Rogers* of Fighter Squadron Seventeen in the Pacific Theater during World War Two. While nursing his crippled F-4U Corsair he was jumped by a pair of Zeros. He downed one before his Corsair was stricken. His final words were "Remember me with the Jolly Rogers." Ens. Ernie was posthumously awarded the Navy Cross. His remains were recovered many years later, after the disestablishment of his former squadron. At the request and with the consent of his family, Ens. Ernie's skull and femurs were encased in glass and presented to the *Jolly Rogers* of VF-84. With the approach of VF-84's disestablishment, Ens. Ernie was transferred to VF-103 on 29 September 1995. (Author)

F-14B-150-GR, **163217**
Left: The *Ghostriders* of VF-142 were disestablished 30 April 1995. Their last cruise was made with CVW-7 embarked aboard the USS George Washington (CVN-73). This example was photographed 30 June 1992 being loaded by crane aboard CVN-73 in preparation for the vessel's commissioning ceremony on 4 July. This Tomcat was delivered in January 1989. Following service with VF-142 it was transferred to the *Ghostriders'* sister squadron, VF-143. By February 1998 it was assigned to VF-103. (USN/PH2 John Sokolowski)

F-14B-150-GR, **163218**
Left: The *Pukin' Dogs* of VF-143 made their first cruise aboard the USS George Washington (CVN-73) 20 May to 18 November 1994. This example was photographed at NAS Fallon as CVW-7 was participating in integrated carrier air wing training for the cruise. This F-14B was delivered in February 1989. In February 1998 and following service with VF-101, this Tomcat was assigned to CFWL. (Grove)

F-14B-150-GR, **163219**
Right: This VF-103 Boss bird was photographed at NAS Fallon in July 1993, as the *Sluggers* and CVW-7 were preparing for a 1994 Med Cruise aboard the USS Saratoga (CV-60). This F-14B was delivered to VF-142 in March 1989. It later served with VF-101 before being transferred to VF-103. By June 1997 it was reassigned to VF-101. (Grove)

F-14B-150-GR, **163220**
Right: Photographed on 23 March 1996, following its 1995 cruise to the Adriatic Sea, this *Diamondbacks* F-14B displays no markings indicative of its role in *Operation Deliberate Force.* During this deployment, VF-102 conducted air strikes against Bosnian Serb ground targets. This Tomcat was delivered to VF-143 in March 1989. It was transferred to VF-102 in 1994 and remained with that squadron until transfer to NAS Jacksonville. As of February 1998 it was assigned to VF-143. (JEM Slides)

F-14B-150-GR, **163221**(Upgrade)
Below: The *Bedevilers* of VF-74 were disestablished 30 April 1994. This F-14B was photographed on 3 November 1993 at Nellis AFB where it was flying as Red Air at Red Flag 94-1. One of the squadron's final missions was providing adversaries for the fleet and USAF. This example had a special camouflage scheme suitable to the squadron's new role. Delivered to VF-74 in April 1989, following the squadrons disestablishment it was transferred to VF-101. This Tomcat was later delivered to Grumman's St. Augustine facility for avionics modifications emerging as an F-14B (Upgrade). As of February 1998 it was assigned to VF-102. (Tunney)

F-14B-150-GR, **163222**(Upgrade)
Left: The *Grim Reapers* of VF-101 are not only tasked with training the fleet's Tomcat crews, they also provide crews and Tomcats for air show performances throughout North America. These demonstrations are awesome and are now even more so with the introduction of the F-14B and F-14D to the inventory. This example streams dual roster tails during a performance at NATC Patuxent River on 22 May 1993. Delivered to VF-211 in May 1989, it was transferred to VX-9 where it was flown to test the F-14B avionics upgrade. As of February 1998 this Tomcat was flying with VF-11. (Roop)

F-14B-150-GR, **163223**(Upgrade)
Left: Uploaded with a single AIM-54 Phoenix, this Naval Weapons Test Squadron Tomcat is the prototype for the F-14B Avionics Upgrade. It was photographed 14 December 1995. Initially delivered to VF-24 in May 1989, it will most likely spend the remainder of its service life devoted to test and evaluation flights. (USN via Kaston)

F-14B-150-GR, **163224**(Upgrade)
Below: This F-14B was delivered to VF-24 in June 1989. When the *Fighting Renegades* transitioned back to the F-14A, this example was transferred to VF-142. It was photographed at NAS Fallon during October 1993. It has since served with VF-101 and VF-102. As of February 1998 it was assigned to VF-103. (Grove)

F-14B-150-GR, **163225**
Right: Delivered to VF-24 in July 1989, this F-14B later served with VF-101, VF-103, and NATC. It was photographed in VF-103 markings preparing to launch from NAS Oceana on 20 October 1995. As of February 1998 it was still assigned to CFWL. (Author)

F-14B-150-GR, **163226**(Upgrade)
Below: The *Fighting Renegades* of VF-24 were equipped with the F-14B from 1989 to 1992. Due to an insufficient number of F-14Bs, the squadron transitioned back to the F-14A. This example was photographed taking off at NAS Miramar in October 1991. Following a brief assignment with VF-101 It was transferred to VX-9 Det. Point Mugu, and has been involved in testing avionics upgrades for the F-14B. (Author)

F-14B-150-GR, **163227**(Upgrade)
Right: The *Fighting Checkmates* of VF-211 deployed twice with the F-14B. Both cruises were aboard the USS Nimitz (CVN-68). As with its sister squadron VF-24, VF-211 transitioned back to the F-14A in 1992. This F-14B was photographed flying from NAS Miramar with a doomsday loadout of six AIM-54 Phoenix missiles. It was delivered to VF-211 in August 1989, and later served with VF-101 and VF-103. As of February 1998 it was assigned to VF-102. (USN)

F-14B-150-GR, **163228**(Upgrade)
Left: This VF-101 F-14B was photographed 25 October 1993 at NAS Oceana. It was initially delivered to VF-24 in September 1989. When that squadron transitioned back to the F-14A, this example was transferred to VF-101. Following its upgrade modifications it was assigned to VF-11. (Kaminski)

F-14B-150-GR, **163229**
Above: Delivered in September 1989, this VF-211 F-14B was photographed shortly before the *Fighting Checkmates*' 1991 WestPac/Indian Ocean cruise aboard the USS Nimitz (CVN-68). This example was later transferred to VF-101 where it was assigned as of February 1998. (Roth)

F-14B-155-GR, **163407**
Left: The orange and blue TACTS pods represent the only color on this *Pukin Dogs* Tomcat when it was photographed at NAS Fallon in October 1993. This F-14B was delivered in October 1989. It served with VF-24 prior to its assignment to VF-143. As of February 1998 it was still flying with the *Pukin' Dogs*. (Grove)

F-14B-155-GR, **163408**(Upgrade)
Above: The *Bedevilers* of VF-74 flew their final sortie on 15 April 1994, ending 50 years of flight operations. Following disestablishment on 30 April 1994, this F-14B was transferred to VF-101. On 18 January 1996, this Tomcat, crewed by Lt. Dane Dobbs and Lt. Jim Skarbek, made the first arrested landing aboard the USS John C. Stennis (CVN-74). By September 1997 this F-14B was assigned to VF-11. (Grove)

F-14B-155-GR, **163409**
Right: This VF-102 Tomcat was photographed in September 1995 following the *Diamondbacks'* participation in *Operation Deliberate Force.* This F-14B was delivered to VF-24 in December 1989 and has served with VF-24, VF-143, and VF-102. As of February 1998 it was again assigned to VF-143. (Author)

F-14B-155-GR, **163410**(Upgrade)
Right: Delivered to VF-211 in January 1990, this F-14B was photographed at NAS Fallon in August 1992. Of interest is the BRU-32 bomb rack attached to the weapons rail. This Tomcat later served with VF-101 prior to being transferred to VF-143. In June 1997 it was photographed in VF-11 markings following the *Red Rippers'* transition from the F-14D to the F-14B. (Grove)

F-14B-155-GR, **163411**
Above: The 595th Tomcat and last F-14B was stricken on 15 March 1993. Fortunately, it was captured on film in February 1993, displaying VF-101 boss bird markings. Initially delivered to VF-24, it was transferred to VF-101 when the *Fighting Renegades* transitioned back to the F-14A in 1992. (Grove)

F-14D-160-GR, **163412**
Left: Unofficially dubbed the "Super Tomcat", the F-14D retains the improvements incorporated in the F-14B, plus new avionics including the AN/APG-71 radar, AN/ALR-67 Radar Warning Receiver, AN/ALQ-165 Advanced Self Projection Jammer, Joint Tactical Information Distribution System, and a new Infrared Search and Track set. This example was delivered to NATC in May 1990, and has carried out the test and evaluation mission there ever since. Most recently it has been used to test the Tomcat's new Digital Flight Control System. (Pugh)

F-14D-160-GR, **163413**
Left: The second F-14D was delivered to VX-4 at Point Mugu in June 1990. Following the disestablishment of the *Evaluators*, it was reassigned to VX-9, Det. Point Mugu. As of February 1998 it was assigned to VF-213. (Author)

F-14D pilot's cockpit.

F-14D RIO's cockpit.

F-14D-160-GR, **163414**
Left: This F-14D wears a variation of the VX-4 tail markings, which appear to be more politically correct than the Playboy Bunny. It was photographed at Point Mugu in June 1993. This example was delivered to Point Mugu in June 1990. The *Evaluators* took the Super Tomcat aboard the USS Nimitz (CVN-68) for carrier trials in August 1990. As of February 1998 this Tomcat was assigned to VF-213. (Roth)

F-14D-160-GR, **163415**
Above: Wearing new tail markings, this Navy Weapons Test Squadron NF-14D was photographed 14 December 1995. This example has been modified with instrumentation and separation cameras. It was still assigned to Point Mugu as of February 1998. (Vasquez)

F-14D-160-GR, **163416**
Left: This NF-14D is a Boss bird of sorts for the Naval Air Weapons Center-Weapons Division. It was delivered in August 1990 and is the 600th Tomcat built for the Navy. It is dedicated to the test mission at Point Mugu and has been permanently modified to carry instrumentation, telemetry, and camera gear. Note the missing panel for the refueling probe. This item would occasionally separate during aerial refueling. (Author)

F-14D-160-GR, **163417**
Right: Photographed at Patuxent River, NATC, on 25 May 1991, this NF-14D was on static display during the open house. Patuxent River and Point Mugu continue to explore the capabilities of the last of the Grumman cats. This Tomcat remained at Patuxent River through February 1998. (Roop)

F-14D-160-GR, **163418**
Right: The *Gunfighters* of VF-124, the Pacific Fleet FRS, took delivery of this F-14D, its first, on 18 October 1990. On 2 October 1992, the squadron flew four Super Tomcats to the USS Nimitz (CVN-68) for the F-14D's first fleet carrier qualifications. This example was transferred to VF-11, which moved to NAS Miramar from Oceana to transition to the F-14D. Following service with VF-101 it was transferred to CFWL where it remained through February 1998. (Puzzullo)

F-14D-165-GR, **163893**
Below: This example was delivered to VF-124 in October 1990. It later served with VF-101, Det. A before transfer to VF-31 in September 1995. Resting on the NAS Miramar ramp, this VF-31 CAG bird was photographed 23 October 1995. The *Tomcatters* and their sister squadron, VF-11, transitioned to the F-14D from 1 March 1992 through 1 June 1992. Felix's first flight on a VF-31 F-14D took place on 9 July 1992. As of This example was delivered to VF-124 in October 1990. It later served with VF-101, Det. A before transfer to VF-31 in September 1995. As of February 1998 it was assigned to VF-213. (Author)

F-14D-165-GR, **163894**
Left: The *Gunfighters* of VF-124 took delivery of this F-14D in October 1990. In 1994 it was transferred to VF-101, Det. A. Initially, VF-51 and VF-111 were scheduled to become the first operational Super Tomcat squadrons. However, neither squadron made the conversion. Instead, both were disestablished on 16 February 1995. In 1992, VF-11 and VF-31 became the first operational F-14D squadrons. This example, in VF-124 markings, was photographed in February 1991. It served briefly with VF-101 and as of September 1997 was flying with VF-2. (Puzzullo)

F-14D-165-GR, **163895**
Left: This Tomcat was delivered to VF-124 15 November 1990. It was photographed at Nellis AFB, 24 April 1998 while participating in Green-Flag 98-2. (Author)

F-14D-165-GR, **163896**
Below: This VF-124 Super Tomcat was photographed at NAS Miramar 2 May 1992. Delivered to the squadron in June 1991, it was transferred to VF-101, Det. A in September 1994 following the disestablishment of VF-124. As of February 1998 it was assigned to VF-101. (Van Aken)

F-14D-165-GR, **163897**
Right: Compare the ejection seats of this F-14D with those of the F-14A parked next to it. The F-14A is equipped with Martin-Baker GRU-7 ejection seats. The F-14D is equipped with Martin-Baker MB-14 (NACES) ejection seats. This VF-124 Tomcat was photographed at NAS Miramar in January 1991. It was delivered to the squadron 12 December 1990. Following the disestablishment of the *Gunfighters,* this Tomcat was transferred to VF-101, Det. A. As of September 1997 it was assigned to VF-2. (Roth)

F-14D-165-GR, **163898**
Below: On 27 March 1994, the *Bounty Hunters* of VF-2 were deployed to NAS Fallon for Air Wing strike training. During this Det, the air wing, consisting of one Tomcat and three Hornet squadrons, achieved an impressive 22 to 1 kill ratio against Fallon's adversaries. This example was delivered to VF-124 in December 1990. It was transferred to VF-2 in 1994 remaining there through February 1998. (Grove)

F-14D-165-GR, **163899**
Right: This F-14D was photographed at Point Mugu while assigned to the Weapons Division of NAWC on 24 October 1995. Delivered to VF-124 on 26 January 1991, it later served with NAWC-AD from 1993 until transferred to VF-101 in March 1996. (Author)

F-14D-165-GR, **163900**
Above: This *Red Rippers* Tomcat was photographed returning from NAS Miramar from a practice air-to-ground sortie on 23 October 1995. Of interest are a pair of ITERs mounted on the forward weapons rails. This F-14D was delivered to VF-124 on 24 March 1991. It was transferred to VF-11 in 1993. The *Red Rippers* have returned to NAS Oceana and have transitioned to the F-14B. This example was assigned to VF-31 as of February 1998. (Author)

F-14D-165-GR, **163901**
Left: Delivered to VF-124 in March 1991, this F-14D was photographed in VF-31 livery at NAS Miramar in January 1994. The *Tomcatters* are assigned to CVW-14, and made their first cruise aboard the USS Carl Vinson (CVN-70) 18 February to 15 August 1994. As of May 1997 this example was assigned to VF-31 and based at NAS Oceana. (Grove)

F-14D-165-GR, **163902**
Left: A flying pencil no more, the red tails and black nose had become a thing of the past by the time this VF-31 Super Tomcat was photographed at NAS Miramar on 24 April 1993. This example was delivered 31 May 1991. It served with VF-124 before being transferred to VF-31 in 1993. It remained with the *Tomcatters* through February 1998. (Tunney)

F-14D-165-GR, **163903**
Above: Although this VF-124 Super Tomcat does not carry the 00 side number normally associated with a CAG or Boss bird, on the canopy rail it displays the CO's name and that of RAdm. Anselmo, who was the Commander of COMFITEAWWINGPAC. This F-14D was delivered to VF-124 in June 1991. It was transferred to VF-2 in August 1993. By February 1998 it was assigned to VF-213. (Author)

F-14D-165-GR, **163904**
Right: The *Red Rippers* are the Navy's second-oldest continuously active fighter squadron. This VF-11 CAG bird was photographed on 23 October 1995 following the squadron's participation in a "Pineapple Cruise" to Hawaii. The cruise commemorated the 50th anniversary of the end of World War Two. The name SOY, under the RIO's canopy, stands for Sailor of the Year. His name appears on the canopy rail. This example was delivered to VF-124 in July 1991 and later transferred to VF-11 in 1992. As of February 1998 it was serving with VF-31. (Author)

F-14D-170-GR, **164340**
Right: This Super Tomcat, VF-31's CAG bird, was photographed at Point Mugu in September 1994. It was stricken on 13 January 1995 when it collided with F-14D(R) 159635, which at the time was also assigned to VF-31. (Roth)

F-14D-170-GR, **164341**
Left: This Super Tomcat was delivered to VF-124 on 22 August 1991. It was transferred to VF-11 in June 1992 and was one of the last F-14Ds assigned to the *Red Rippers* prior to that squadron's transition to the F-14B. As of February 1998 this Tomcat was assigned to VF-31. (Tunney)

F-14D-170-GR, **164342**
Left: This VF-31 Super Tomcat had been prepped for the wash rack when it was photographed in May 1992. Compare this example with 163902 and note the subtle changes in tail markings. This F-14D has been assigned to VF-124, VF-31, and VF-101, Det. A. As of February 1998 it was assigned to VF-101. (Grove)

F-14D-170-GR, **164343**
Below: Photographed at NAS Miramar on 5 December 1992, this Super Tomcat was delivered to VF-124 on 20 September 1991, and transferred to VF-11 in 1992. As of February 1998 it was assigned to VF-101. (Van Aken)

F-14D-170-GR, **164344**
Right: Looking very much like an F-14B, this VF-31 F-14D was photographed 8 August 1992. The principal external differences between the F-14D and its predecessor, the F-14B, are the NACES ejection seats and the side-by-side chin mounted TCS – Infrared Search and Tracking Set (IRST). This example was delivered to VF-124 in October 1991. It was transferred to VF-31 in 1992. It was still assigned as of February 1998. (Snyder)

F-14D-170-GR, **164345**
Below: Delivered to VF-124 in December 1991, this Super Tomcat was transferred to VF-11 in June 1992 and photographed in their markings in February 1993. On 30 November 1996, the *Red Rippers* became the first F-14D squadron to drop bombs at night using Night Vision Goggles (NVGs). As of February 1998, this Super Tomcat was assigned to VF-31. (Grove)

F-14D-170-GR, **164346**
Right: The *Tomcatters* of VF-31 were conducting integrated air wing work-ups when this Super Tomcat was photographed at NAS Fallon in September 1993. The air wing, CVW-14, was preparing for a WestPac cruise aboard the USS Carl Vinson (CVN-70). The cruise marked VF-31's first trip to the western Pacific since its 1972 Vietnam cruise. This F-14D was delivered to VX-4 in March 1992. It was transferred to VF-31 in July. As of February 1998 it was still assigned to the *Tomcatters*. (Grove)

F-14D-170-GR, **164347**
Left: This VF-11 Super Tomcat, minus its engines, was photographed at NAS Miramar in April 1992. These same powerplants make the F-14D truly super, providing it with a marked increase in thrust and the ability to launch from a carrier without the use of afterburners. This example was delivered to VF-124 in January 1992 and reassigned to VF-11 in June. As of February 1998 it was assigned to VF-101. (Grove)

F-14D-170-GR, **164348**
Above: Delivered to VF-124 early in 1992, this Super Tomcat was transferred to VF-31 in June. It was serving as the *Tomcatters* CAG bird when it was photographed at Roswell Industrial Air Center, New Mexico while participating in *Roving Sands '97*. (Author)

F-14D-170-GR, **164349**
Left: In 1991, the *Red Rippers* of VF-11 moved from NAS Oceana to NAS Miramar. Once there, they transitioned to the F-14D, becoming the first operational Super Tomcat squadron. Delivered to VF-124 in February 1992, this example was photographed at NAS Miramar in May 1992, following its delivery to VF-11. As of February 1998 it was assigned to VF-2. (Grove)

F-14D-170-GR, **164350**
Right: Callsign "Bandwagon 205", this Tomcat made the first flight by a VF-31 F-14D on 9 July 1992. It was delivered to VF-124 in February 1992. Following service with VF-31 it was transferred to VF-2 where it remains assigned through February 1998. This example was photographed at NAS Miramar 5 December 1992. (Roth)

F-14D-170-GR, **164351**
Right: This VF-2 Tomcat was initiallyy delivered to VF-124 on 24 March 1992. It was photographed taking off from Nellis AFB, 24 April 1998 while participating in Green Flag 98-2. The shape carried on the ITER is a BDU-57B Laser Guided Target Round. (Author)

F-14D-170-GR, **164599**
Below: Armed with CATM-7M and CATM-9M practice air-to-air missiles, this VX-4 Super Tomcat was photographed at Point Mugu in July 1994. The *Evaluators* have been heavily involved in testing the AIM-7M, AIM-9M and the AIM-54C+. Due to its test mission the squadron has a direct impact on tactics and the acquisition of fighter weapons. VX-4 was disestablished on 30 September 1994 and its assets were reassigned to VX-9 Det. Point Mugu. This Super Tomcat is currently assigned to VX-9 at Point Mugu. (Roth)

F-14D-170-GR, **164600**
Above: All fleet Tomcat training was handed over to VF-101 when the *Gunfighters* of VF-124 stood down in September 1994. This example, a VF-101 Det. Miramar Super Tomcat, was photographed at Miramar on 16 February 1994. This Tomcat was assigned to VF-31 as of February 1998. (Van Der Lugt)

F-14D-170-GR, **164601**
Left: This VF-101 Det. Miramar Super Tomcat was photographed wearing a temporary water soluble paint scheme. This example was delivered to VF-124 17 April 1992. It was transferred to VF-101 in September 1994 and was assigned there as of February 1998. (USN via Kaston)

F-14D-170-GR. **164602**
Left: Photographed at the NAS Miramar Open House on 14 August 1992, this VF-124 Super Tomcat was delivered 1 May 1992. Following service with VF-2 it was reassigned to VF-213 when the *Black Lions* converted to F-14Ds in late-1997. (Van Aken)

F-14D-170-GR, **164603**
Above: This Super Tomcat was photographed in June 1995 at the London Ontario Canada, International Air Show. It was delivered to VF-124 on 29 May 1992 and reassigned to VF-2 in June 1993. By February 1998 it was flying with VF-213. (Author)

F-14D-170-GR, **164604**
Right and below: Two photographs of the 'Black Bunny", excuse me, "Vandy One", are not sufficient to do justice to this eye-catching paint scheme. For more than twenty years, VX-9 and its predecessor, VX-4, have selected one of its assigned fighter aircraft, to wear these distinctive markings. This Super Tomcat, the last Grumman Cat constructed, was delivered on 10 July 1992. As of February 1998 it was assigned to VX-9. (Vasquez and Roth)

Tail Colors

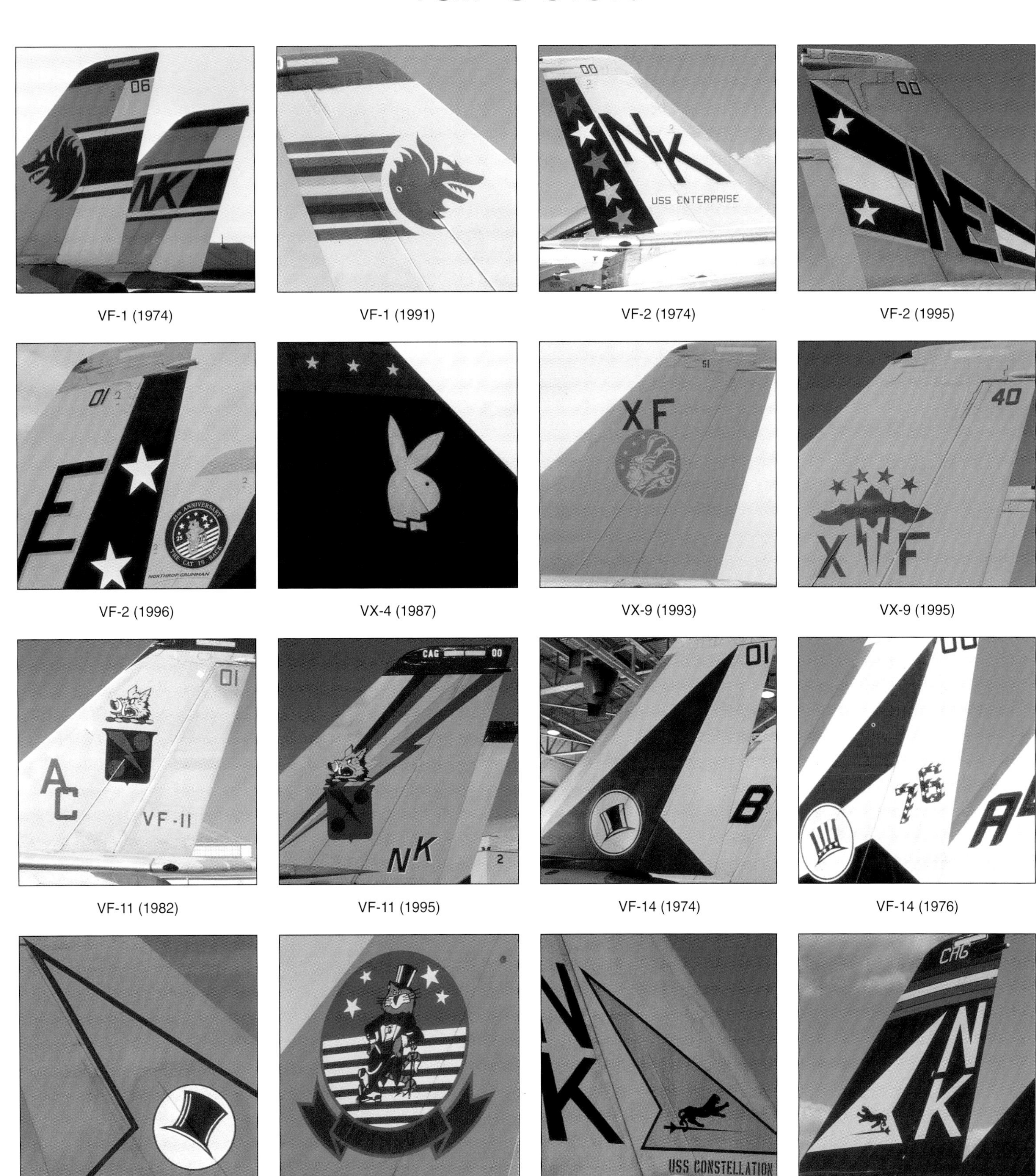

VF-1 (1974) | VF-1 (1991) | VF-2 (1974) | VF-2 (1995)

VF-2 (1996) | VX-4 (1987) | VX-9 (1993) | VX-9 (1995)

VF-11 (1982) | VF-11 (1995) | VF-14 (1974) | VF-14 (1976)

VF-14 (1994) | VF-14 (1997) | VF-21 (1986) | VF-21 (1991)

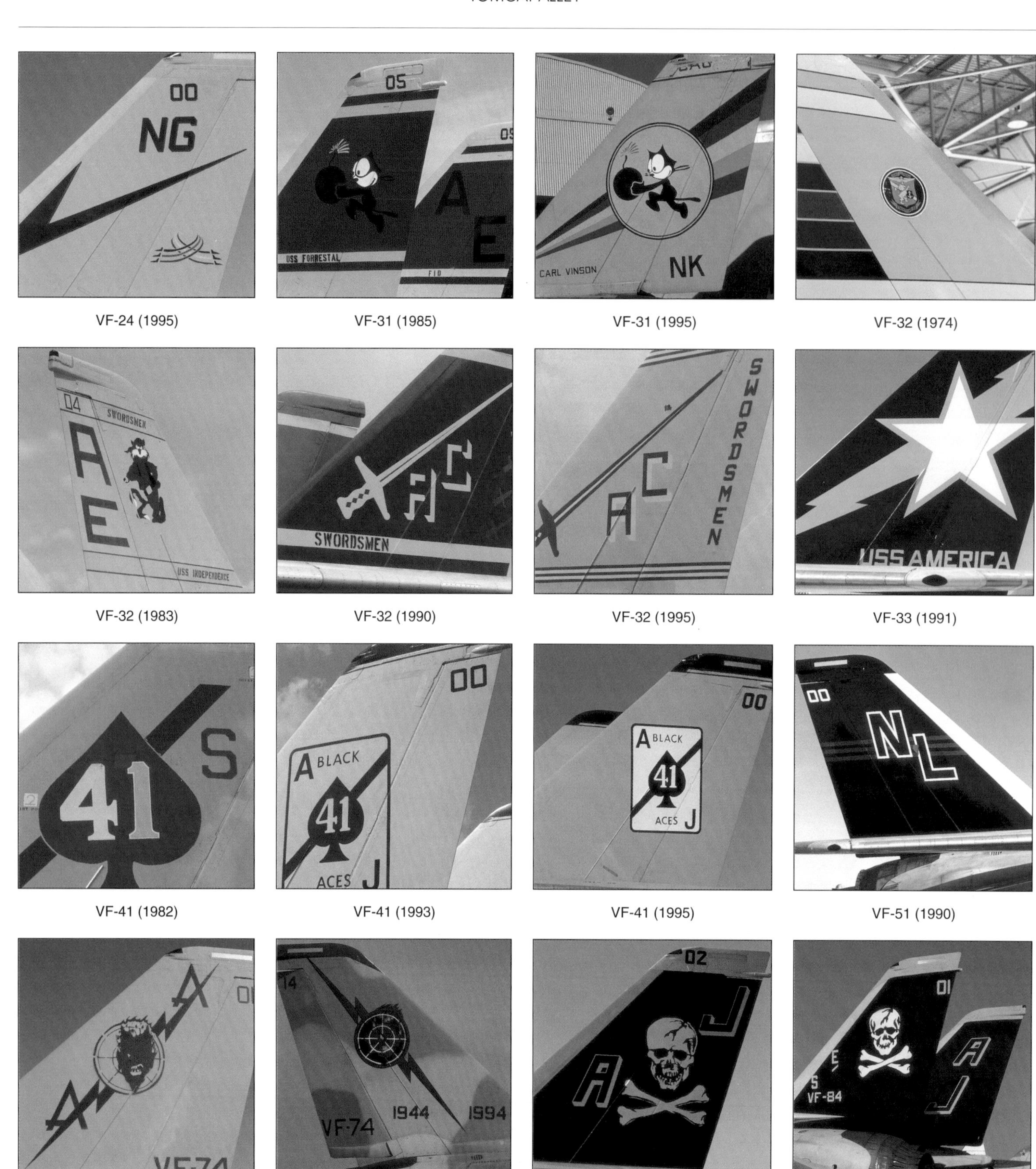

VF-24 (1995)

VF-31 (1985)

VF-31 (1995)

VF-32 (1974)

VF-32 (1983)

VF-32 (1990)

VF-32 (1995)

VF-33 (1991)

VF-41 (1982)

VF-41 (1993)

VF-41 (1995)

VF-51 (1990)

VF-74 (1991)

VF-74 (1994)

VF-84 (1978)

VF-84 (1989)

VF-84 (1991)

VF-84 (1994)

VF-101 (1976)

VF-101 (1986)

VF-101 (1988)

VF-101 (1995)

VF-101 (1995)

VF-101 (1995: VF-1 commem.)

VF-101 (1996: VF-21 commem.)

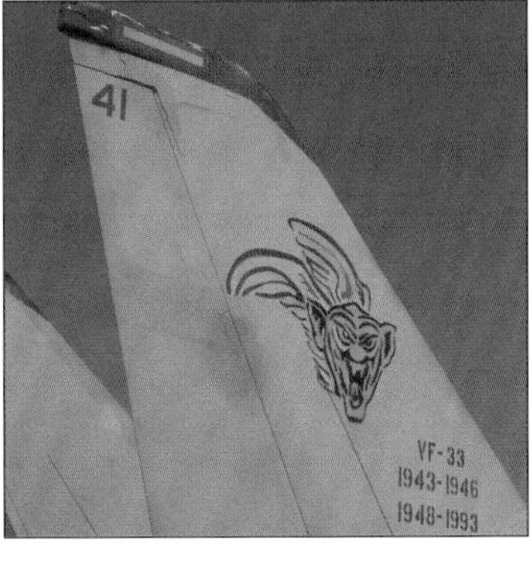

VF-101 (1995: VF-33 commem.)

VF-101 (1995: VF-74 commem.)

VF-101 (1995: VF-114 commem.)

VF-101 (1995: VF-142 commem.)

VF-101 (1995: VF-111 commem.)

VF-102 (1988)

VF-102 (1994)

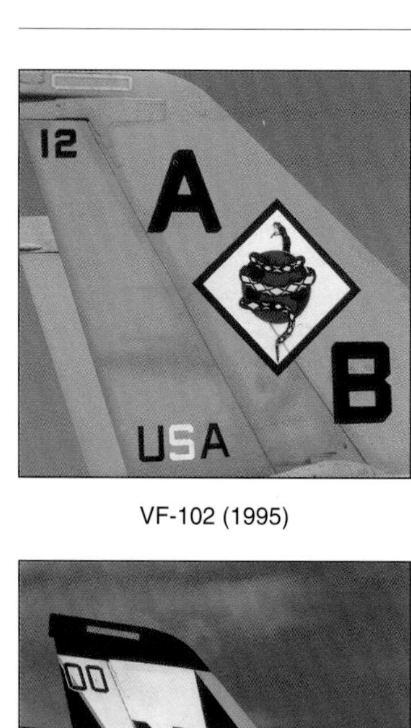

VF-102 (1995)

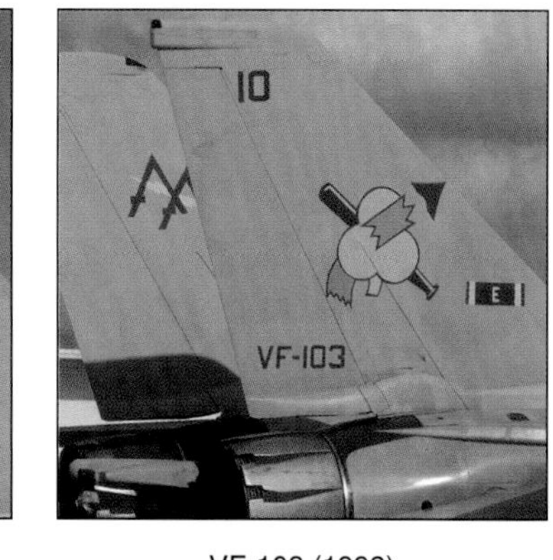

VF-103 (1992)

VF-103 (1995)

VF-103 (1995)

VF-111 (1995)

VF-114 (1976)

VF-114 (1992)

VF-124 (1986)

VF-124 (1987)

VF-124 (1989)

VF-124 (1990)

VF-142 (1975)

VF-143 (1975)

VF-154 (1988)

VF-154 (1991)

VF-191 (1987)

VF-194 (1988)

VF-201 (1987)

VF-201 (1993)

VF-202 (1987)

VF-211 (1979)

VF-211 (1995)

VF-213 (1976)

VF-213 (1995)

VF-302 (1986)

TOPGUN (1991)

TOPGUN (1992)

PMTC (1989)

PMTC (1992)

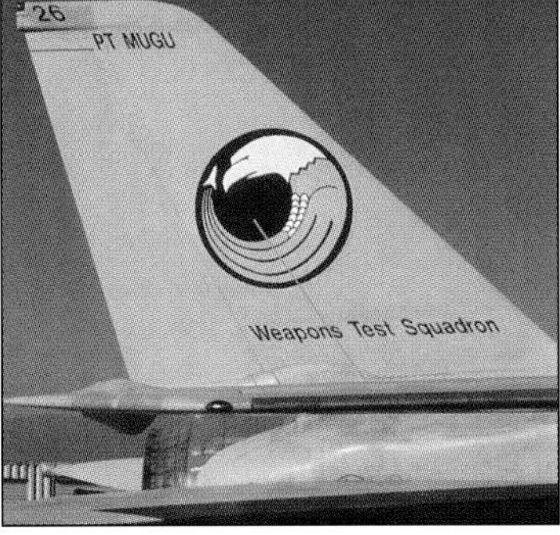

PMTC (1995)

NAWC (1995)

NASA

Patches

VF-1 VF-1 VF-1 VF-1

VF-2 VF-2 VF-2 VF-2

VX-4 VX-4 VX-4 VX-4

VX-4 VX-9 VF-11 VF-14

VF-14 VF-14 VF-14 VF-14

VF-21 VF-24 VF-24 VF-31

VF-31 VF-31 VF-32 VF-32

VF-32 VF-32 VF-32 VF-33

VF-33 | VF-41 | VF-41 | VF-41

VF-51 | VF-74 | VF-74 | VF-74

VF-84 | VF-84 | VF-101 | VF-101

VF-101 | VF-102 | VF-102 | VF-102

VF-103 VF-103 VF-111 VF-111

VF-111 VF-111 VF-114 VF-114

VF-114 VF-114 VF-114 VF-114

VF-124 VF-124 VF-124 VF-142

VF-142 VF-143 VF-143 VF-143

VF-154 VF-154 VF-154 VF-154

VF-191 VF-194 VF-201 VF-201

VF-202 VF-202 VF-211 VF-211

VF-211 | VF-213 | VF-213 | VF-213

VF-213 | VF-301 | VF-301 | VF-301

VF-301 | VF-302 | VF-1485 | OAG

THE EYES OF THE EAGLE
EAGLE SHOOTER
WEAPONS TEST SQUADRON
POINT MUGU
CVW 3
DESERT STORM
1991
RED SEA
HASTA LA VISTA, BABY
NAS MIRAMAR
FIGHTERTOWN USA
GONZO STATION-87
ACK, THTHPT BABY..!
FFARP
LIFE IS HARD
THEN YOU DIE
FIGHTER WING PACIFIC
JANUARY
1997
VENI, VIDI, VACATE
89
WESTPAC 89
INDIAN OCEAN CRUISE 1985
NO FUN
SEMI-FUN
MAX FUN
B.F.D. FUN METER
64
CVW-14 "CONNIE"
BATTLE FLEX DECK
TOMCAT
ANYTIME BABY
F-14 TOMCAT
ANYTIME, BABY. . . !
ANYTIME, BABY...!
A+
F-14A PLUS
TOMCAT
GO AHEAD...MAKE MY DAY
SUPER TOMCAT
SUPER

AAACK!
ANYTIME, SENATOR...!
TOMCAT
PEEPING TOM
TARPS
SUPER TOMCAT
F-14D
SUPER TOMCAT
F-14
GO AHEAD... MAKE MY DAY
F-14
GO AHEAD...MAKE MY DAY
TOMCAT TWEAKERS
TOMCAT TWEAKERS
PET ME TODAY... I GAVE!
GRUMMAN DONOR
WORLDWIDE RECONNAISSANCE AIR MEET
1988
A+
F-14A PLUS
500 HRS
1500 HRS
2000 HRS
2500 HRS
100 TIMES, BABY...!
200 TIMES. BABY...!
300 TIMES. BABY...!

100 DARK ONES, BABY!
R.I.O., BABY!
F-14
F-14 VSTFE...
SMOOTH, BABY!
LANTIRN
TOMCAT
HUGHES
APG 71
RADAR
AWG-9 PHOENIX
WEAPON SYSTEM
The Evaluators
AIRTEVRON NINE DET POINT MUGU
NAVAL AIR SYSTEMS
COMMAND
NAVAL AIR TEST CENTER
STRIKE AIRCRAFT TEST
TOMCAT
STRIKE FIGHTER
WEAPONS SCHOOL
PMTC FLIGHT TEST
AIRCRAFT
SYSTEMS
FLIGHT TEST
NAWC
Weapons Division
F-14 WSIC FACILITY
U.S. NAVY PT. MUGU
FLIGHT TEST
TOMCAT

Appendices

TOMCAT EQUIPPED SQUADRONS

DEPLOYABLE SQUADRONS

VF-1	*Wolfpack*	Disestablished 1 October 1993
VF-2	*Bounty Hunters*	NAS Oceana
VF-11	*Red Rippers*	NAS Oceana
VF-14	*Tophatters*	NAS Oceana
VF-21	*Freelancers*	Disestablished 31 January 1996
VF-24	*Fighting Renegades*	Disestablished 31 August 1996
VF-31	*Tomcatters*	NAS Oceana
VF-32	*Swordsmen*	NAS Oceana
VF-33	*Tarsiers/Starfighters*	Disestablished 1 October 1993
VF-41	*Black Aces*	NAS Oceana
VF-51	*Screaming Eagles*	Disestablished 31 March 1995
VF-74	*Bedevilers*	Disestablished 30 April 1994
VF-84	*Jolly Rogers*	Disestablished 1 October 1995
VF-102	*Diamondbacks*	NAS Oceana
VF-103	*Sluggers/Jolly Rogers*	NAS Oceana
VF-111	*Sundowners*	Disestablished 31 March 1995
VF-114	*Aardvarks*	Disestablished 30 April 1993
VF-142	*Ghostriders*	Disestablished 30 April 1995
VF-143	*Pukin' Dogs*	NAS Oceana
VF-154	*Vigilantes/Black Knights*	NAF Atsugi, Japan
VF-191	*Satan's Kittens/Angels*	Disestablished 30 April 1988
VF-194	*Red Lightnings*	Disestablished 30 April 1988
VF-211	*Flying Checkmates*	NAS Oceana
VF-213	*Black Lions*	NAS Oceana

RESERVE SQUADRONS

VF-201	*Hunters*	NAS Fort Worth, JRB
VF-202	*Superheats*	Disestablished 1 October 1994
VF-301	*Devil's Disciples/Infernos*	Disestablished 31 December 1994
VF-302	*Stallions*	Disestablished 31 December 1994

FLEET READINESS SQUADRONS

VF-101	*Grim Reapers*	NAS Oceana
VF-101	*Grim Reapers* Det. A Miramar	Disestablished
VF-124	*Gunfighters*	Disestablished 30 September 1994

TEST AND EVALUATION

VX-4	*Evaluators*	Disestablished 30 September 1994
VX-9	*Vampires*	NAWS Point Mugu
NADC		Incorporated with NAWC, 2 Jan. 1992
NAWC-WD		NAWS Point Mugu, NAS China Lake
NAWC-AD		NAS Patuxent River,
NAWC Warminster		Closed 30 October 1996
NWTS-PM		Naval Weapons Test Sqdn-Point Mugu
NSAWC		Naval Strike and Air Warfare Center NAS Fallon
NWEF		Kirtland AFB

TOMCATS STORED AT AMARC AS OF 4 AUGUST 1997

Bureau Number	Arrival Date	Storage Code
158622	18 December 1991	AN1K0052
158979	30 August 1990	AN1K0013
158980	09 September 1990	AN1K0015
158986	19 July 1991	AN1K0033
158987	25 October 1990	AN1K0018
158988	18 September 90	AN1K0016
158989	29 August 1991	AN1K0040
158990	26 August 1991	AN1K0039
158991	22 August 1990	AN1K0011
158992	28 February 1991	AN1K0028
158993	25 October 1990	AN1K0019
158994	27 August 1990	AN1K0012
159000	20 August 1990	AN1K0009
159002	17 August 1990	AN1K0008
159003	10 September 1990	AN1K0014
159004	27 September 1991	AN1K0047
159005	24 September 1991	AN1K0044
159006	09 September 1991	AN1K0042
159009	31 July 1991	AN1K0036
159010	05 June 1991	AN1K0031
159013	20 August 1990	AN1K0010
159014	23 July 1991	AN1K0034
159015	04 September 1991	AN1K0041
159016	15 July 1991	AN1K0032
159017	24 September 1991	AN1K0045
159018	30 April 1991	AN1K0030
159020	12 August 1991	AN1K0037
159021	26 September 1991	AN1K0046
159422	26 July 1991	AN1K0035
159423	06 December 1990	AN1K0024
159425	14 August 1991	AN1K0038
159426	29 November 1990	AN1K0021
159427	13 September 1991	AN1K0043
159429	16 January 1992	AN1K0055
159433	15 November 1991	AN1K0049
159434	18 December 1991	AN1K0053
159435	31 October 1991	AN1K0048
159436	19 November 1991	AN1K0050
159437	11 February 1992	AN1K0056
159438	18 March 1992	AN1K0058
159440	05 March 1992	AN1K0057
159442	20 September 1990	AN1K0017
159444	01 April 1992	AN1K0059
159447	17 April 1992	AN1K0060
159449	13 May 1992	AN1K0061
159452	14 October 1994	AN1K0071
159453	29 May 1992	AN1K0062
159454	26 July 1994	AN1K0067
159460	20 November 1990	AN1K0020
159462	05 August 1992	AN1K0063
159466	01 October 1992	AN1K0065

Bureau Number	Arrival Date	Storage Code
159467	05 October 1994	AN1K0070
159468	29 June 1994	AN1K0066
159589	03 December 1991	AN1K0051
159596	12 December 1990	AN1K0025
159597	08 December 1994	AN1K0073
159598	29 November 1990	AN1K0022
159604	01 October 1992	AN1K0064
159608	26 October 1994	AN1K0072
159614	20 December 1991	AN1K0054
159621	02 August 1994	AN1K0068
159624	05 December 1990	AN1K0023
159625	31 January 1991	AN1K0027
159634	17 December 1990	AN1K0026
159833	15 February 1996	AN1K0087
159834	21 March 1991	AN1K0029
159837	07 July 1995	AN1K0079
159849	21 March 1995	AN1K0076
159858	23 August 1995	AN1K0081
159862	21 August 1995	AN1K0080
159863	18 April 1997	AN1K0100
159866	01 May 1996	AN1K0092
160384	31 August 1995	AN1K0084
160387	13 June 1996	AN1K0097
160393	16 April 1996	AN1K0089
160399	12 June 1996	AN1K0096
160402	10 January 1996	AN1K0086
160404	20 April 1996	AN1K0090
160405	25 July 1997	AN1K0105
160406	21 November 1997	AN1K0115
160410	26 April 1995	AN1K0077
160413	26 May 1996	AN1K0094
160655	27 May 1997	AN1K0104
160668	16 March 1995	AN1K0075
160679	04 December 1996	AN1K0099
160680	13 March 1995	AN1K0074
160682	31 May 1996	AN1K0095
160684	26 September 1994	AN1K0069
160697	24 April 1997	AN1K0113
160892	26 April 1995	AN1K0078
160893	14 November 1997	AN1K0114
160897	24 April 1996	AN1K0091
160901	20 March 1996	AN1K0088
160904	31 August 1995	AN1K0083
160908	18 December 1997	AN1K0116
160927	08 September 1995	AN1K0085
160929	23 August 1995	AN1K0082
161137	22 May1996	AN1K0093
161156	16 November 1996	AN1K0098
161161	20 May 1997	AN1K0103
161599	22 April 1997	AN1K00101
161851	22 April 1997	AN1K00102

IRANIAN TOMCATS

Fab No.	Bu. No.	Tail No.	Iranian Tail	Prod. Block
H1	160299	3-1101	3-6001	05
H2	160300	3-1102	3-6002	05
H3	160301	3-1103	3-6003	05
H4	160302	3-1104	3-6004	05
H5	160303	3-1105	3-6005	05
H6	160304	3-1106	3-6006	05
H7	160305	3-1107	3-6007	05
H8	160306	3-1108	3-6008	05
H9	160307	3-1109	3-6009	05
H10	160308	3-1110	3-6010	05
H11	160309	3-1111	3-6011	05
H12	160310	3-1112	3-6012	05
H13*	160311	3-1113	3-6013	05*
H14	160312	3-1114	3-6014	05
H15	160313	3-1115	3-6015	05
H16	160314	3-1116	3-6016	05
H17	160315	3-1117	3-6017	05
H18	160316	3-1118	3-6018	05
H19	160317	3-1119	3-6019	05
H20	160318	3-1120	3-6020	05
H21	160319	3-1121	3-6021	05
H22	160320	3-1122	3-6022	05
H23	160321	3-1123	3-6023	05
H24	160322	3-1124	3-6024	05
H25	160323	3-1125	3-6025	05
H26	160324	3-1126	3-6026	05
H27	160325	3-1127	3-6027	05
H28	160326	3-1128	3-6028	05
H29	160327	3-1129	3-6029	05
H30	160328	3-1130	3-6030	05
H31	160329	3-1131	3-6031	10
H32	160330	3-1132	3-6032	10
H33	160331	3-1133	3-6033	10
H34	160332	3-1134	3-6034	10
H35	160333	3-1135	3-6035	10
H36	160334	3-1136	3-6036	10
H37	160335	3-1137	3-6037	10
H38	160336	3-1138	3-6038	10
H39	160337	3-1139	3-6039	10
H40	160338	3-1140	3-6040	10
H41	160339	3-1141	3-6041	10
H42	160340	3-1142	3-6042	10
H43	160341	3-1143	3-6043	10
H44	160342	3-1144	3-6044	10
H45	160343	3-1145	3-6045	10
H46	160344	3-1146	3-6046	10
H47	160345	3-1147	3-6047	10
H48*	160346	3-1148	3-6048	10*
H49	160347	3-1149	3-6049	10
H50	160348	3-1150	3-6050	10
H51	160349	3-1151	3-6051	10
H52	160350	3-1152	3-6052	10
H53	160351	3-1153	3-6053	10
H54	160352	3-1154	3-6054	10
H55	160353	3-1155	3-6055	10
H56	160354	3 1156	3-6056	10
H57	160355	3-1157	3-6057	10
H58	160356	3-1158	3-6058	10
H59	160357	3-1159	3-6059	10
H60	160358	3-1160	3-6060	10
H61	160359	3-1161	3-6061	10
H62	160360	3-1162	3-6062	10
H63	160361	3-1163	3-6063	15
H64	160362	3-1164	3-6064	15
H65	160363	3-1165	3-6065	15
H66	160364	3-1166	3-6066	15
H67	160365	3-1167	3-6067	15
H68	160366	3-1168	3-6068	15
H69	160367	3-1169	3-6069	15
H70	160368	3-1170	3-6070	15
H71	160369	3-1171	3-6071	15
H72	160370	3-1172	3-6072	15
H73	160371	3-1173	3-6073	15
H74	160372	3-1174	3-6074	15
H75	160373	3-1175	3-6075	15
H76	160374	3-1176	3-6076	15
H77	160375	3-1177	3-6077	15
H78	160376	3-1178	3-6078	15
H79	160377	3-1179	3-6079	15
H80**	160378	3-1180	3-6080	15

*Stricken
**To USN

GLOSSARY

AB	Air Base
ACM	Air Combat Maneuvering
ADC	Air Defense Command
AEW	Airborne Early Warning
AFB	Air Force Base
AGM	Air-to-Ground Missile
AIM	Air Intercept Missile
ACEVAL	Air Combat Evaluation
AIMVAL	Air Intercept Missile Evaluation
AIS	Air Instrumentation System
Alert 5	Readiness condition on the carrier deck in which aircraft are spotted and manned in order to be launched within five minutes.
Alert 15	In the ready room in flightgear able to man-up and launch in fifteen minutes.
AMARC	Aerospace Maintenance and Regeneration Center
AMRAAM	Advanced Medium-Range Air-to-Air Missile (AIM-120)
ANG	Air National Guard
ANGB	Air National Guard Base
AOA	Angle Of Attack
APU	Auxiliary Power Unit
AWACS	Airborne Warning And Control System
BARCAP	Barrier Combat Air Patrol
BIS	Bureau of Inspection and Survey
B/N	Bombardier/Navigator
BuNo	Bureau Number, a serial number assigned to an aircraft by the Navy at its time of manufacture.
BVR	Beyond Visual Range
CAG	Carrier Air Group. Term used to refer to the Commander of a Carrier Air Wing
CAP	Combat Air Patrol
CarQuals	Carrier Qualification Landings
Cat	Catapult
CATM	Captive Air Training Missile
CBU	Cluster Bomb Unit
CFWL	Commander Fighter Wing Atlantic
CV	Aircraft Carrier (Conventional)
CVN	Aircraft Carrier (Nuclear)
CVW	Carrier Air Wing
CVWR	Carrier Air Wing (Reserve)
DACM	Dissimilar Air Combat Maneuvering
DFCS	Digital Flight Control System
DOD	Department Of Defense
ECM	Electronic Counter Measures
EGI	Embedded GPS/INS
FCLP	Field Carrier Landing Practice
FEMS	Fatigue/Engine Monitoring System
FFARP	Flight Fighter Air (combat) Readiness Program
FitWepSchool	Fighter Weapons School – TOPGUN
FLIR	Forward-Looking Infra-Red
FOD	Foreign Object Damage
FRS	Fleet Replacement Squadron. Current term for the RAG
GBU	Guided Bomb Unit
GPS	Global Positioning System
HARM	High-Speed Anti-Radiation Missile
HVU CAP	High-Value Unit Combat Air Patrol
HUD	Head Up Display

INS	Inertial Navigation System
IO	Indian Ocean
IR	Infra-Red
ITALD	Improved Tactical Air Launched Decoy
ITER	Improved Triple Ejector Rack
JRB	Joint Reserve Base
JTIDS	Joint Tactical Information Distribution System
LANTIRN	Low-Altitude, Navigation and Targeting, Infra-Red for Night
LGB	Laser Guided Bomb
LSO	Landing Signal Officer
LTS	LANTIRN Targeting System
MAGR	Miniaturized Airborne Receiver
MCAS	Marine Corps Air Station
MFD	Multi-Function Displays
MiGCAP	Combat Air Patrol to protect against enemy aircraft
MiGSweep	Mission to eliminate enemy aircraft
NACES	Navy Aircrew Common Ejection Seat
NAF	Naval Air Facility
NADC	Naval Air Development Center
NADEP	Naval Air Depot
NARF	Naval Air Rework Facility
NAS	Naval Air Station
NATC	Naval Air Test Center
NAWC	Naval Air Weapons Center
NAWC-AD	Naval Air Weapons Center-Aircraft Division
NAWC-WD	Naval Air Weapons Center-Weapons Division
NFO	Naval Flight Officer
NM	Nautical Miles
NMC	Naval Missile Center
NSAWC	Naval Strike and Air Warfare Center
NWEF	Naval Wepaons Evaluation Facility
NWFS	Naval Weapons Fighter School
NWTS-PM	Naval Weapons Test Squadron-Point Mugu
NVG	Night Vision Goggles
PAVE	Precision Avionics Vectoring Equipment
PGM	Precision Guided Munitions
RAF	Royal Air Force
RN	Royal Navy
RDT&E	Research, Development, Test and Evaluation
ResCAP	Rescue Combat Air Patrol
RIO	Radar Intercept Officer
RWR	Radar Warning Receiver
SAM	Surface-to-Air Missile
SAR	Search And Rescue
SARDIP	Stricken Aircraft Reclamation & Disposal In Place
SDLM	Standard Depot Level Maintenance
SFARP	Strike Fighter Advanced Readiness Program
Sortie	A single mission flown by one aircraft
Stricken	Removed from U.S. Navy inventory
TACTS	Tactical Aircrew Combat Training System
TALD	Tactical Air-Launched Decoy
TARCAP	Target Combat Air Patrol
TARPS	Tactical Air Reconnaissance Pod System
TARPS DI	Tactical Air Reconnaissance Pod System Digital
TCS	Television Camera System

TID	Tactical Information Display
Trap	Arrested landing aboard a carrier
USAF	United States Air Force
USAFE	United States Air Force Europe
USMC	United States Marine Corps
USN	United States Navy
VA	Navy Light Attack Squadron
VAQ	Navy Electronic Warfare Squadron
VAW	Navy Airborne Early Warning Squadron
VF	Navy Fighter Squadron
VFA	Navy Fighter and Attack Squadron
VMA	Marine Attack Squadron
VMA(AW)	Marine Attack Squadron (All Weather)
VMAQ	Marine Electronic Warfare Squadron
VMFA	Marine Fighter/Attack Squadron
VS	Navy Anti-Submarine Squadron
Zone 5	Maximum afterburner (TF-30 engine)

NOTES

NOTES

NOTES

NOTES